Dowsing *Magic*

From water finds to dragon lines

Book 1: "Basics"

Grahame Gardner
© 2012

Additional line drawings by Brian Hartley

First published in 2012 by:

Penwith Press
Cornwall
UK
www.penwithpress.co.uk

ISBN   978-0-9533316-5-9

All rights reserved. For permission to reproduce any part of
this book in any form or media, please contact the publisher.

Printed by TJ International Ltd.
Trecerus Industrial Estate
Padstow, Cornwall
PL28 8RW

Table of Contents

Introduction 4

    My Story 6

Chapter 1 Dowsing Theory 13

    Models 17

Chapter 2 Pendulum Practice 25

    Training our responses 26

    Pre-flight checklist 32

Chapter 3 Historical Context 36

Chapter 4 Working with Rods 46

    The V-rod 47

    The L-rod 50

    Dowsing Practice 51

    The Hydrological Cycle 55

Chapter 5 Lost Objects and Missing People 58

    The Dowsing Detective 61

Chapter 6 Map Dowsing 68

    Techniques 69

Chapter 7 Refining technique 77

    More tools 77

    Bobbers and Wands 79

    The Cameron Aurameter 80

    Depthing and other informational dowsing 81

    Witnesses 87

Chapter 8 Introducing Earth Energies 91

    Definitions and terms 93

    Dowsing the leys 96

    Character of energy leys 99

    The bigger picture 102

    Enter the dragon 104

    Summary 107

Chapter 9 Life, the Universe, and Everything 109

    It's all done with magnets, you know 110

    The European model 114

    A wee bit about geopathic stress 118

    It's a conscious Universe out there! 120

Conclusion 122

Foreword

Grahame Gardner is a complex, fascinating and modest man. Few realise that he has an encyclopaedic knowledge of all aspects of dowsing and has been researching and studying the subject since he was a teenager.

That he manages to combine a busy Dowsing practice with the demanding world of a theatre lighting technician is impressive.

We first met over the boardroom table at a Trustees' meeting of The British Society of Dowsers in 2003. Little did we both know that just a few years later he would be elevated to the President's Chair from where he has guided the Society through a very challenging time – by that I mean the difficult economic climate we have all been experiencing since 2008.

One of the best aspects of being Director of BSD is that I get to spend lots of time with the President and all the 'out of hours' discussions over many glasses of wine have greatly added to my own appreciation of the dowsing world. I have long been encouraging Grahame to write this book, the first of a series, thereby giving everyone the chance to benefit from his vast experience and knowledge. It should be an essential handbook for dowsers everywhere.

John Moss
Director
The British Society of Dowsers March 2012

Introduction

What image do you see in your head when someone mentions the word 'dowser'? Is it a mental picture of a tweed-jacketed gentleman marching across a field clutching a forked twig looking for underground water, rather like the chap in Figure 1? This is the classic image of a water diviner at work, but these days you're much more likely to find the dowser clad in the best high-performance outdoor gear and wielding a nylon V-rod in place of the forked twig.

Figure 1 ©BSD

Perhaps you might also have come across a technician from your local water board using some copper L-rods to pinpoint a leaking water main; and that therapist at the local health shop – haven't you seen her twiddling a pendulum over some charts? Dowsing is more common than you might think. It's always been a bit of a closet activity, but dowsing is finally becoming trendy as people realise what a valuable tool it can be in all walks of life, particularly for enhancing one's intuition and decision-making ability. Dowsing may be an old technology, but it is as relevant today as it was to a thirsty Neolithic nomad in the dry African desert.

Dowsing Magic

Dowsing is an innate human ability. Pretty much anyone can dowse to some extent, unless they are dead set against it themselves. And as I hope to demonstrate, most people are dowsing all the time but aren't aware of it; because dowsing is really about learning to listen to your intuition, and you are more aware of what's going on in your subconscious than you might think.

As well as finding water or other underground features such as gas mains, cables and utilities (and breaks in them), dowsing can be used to locate lost items, pets or people, improve health by identifying food intolerances and allergies, survey archaeological sites in advance of more expensive geophysical surveys, and map the mysterious meridians of 'earth energy' that flow through the landscape, connecting ancient sites like stone circles and old churches along invisible ley lines. In fact, dowsing is useful in all areas of human endeavour, from the practical site dowsing work where the emphasis is on the tangible result, right through to the more esoteric end of the spectrum where a dowser might be employed to clear ghosts from your home and check your aura for geopathic stress. More and more people are rediscovering this ancient art and finding new applications for it in our hectic modern lives.

In this book, we will journey through the world of dowsing, shedding light into some of the dustier corners, dispelling superstitions and demystifying jargon, and learning how to confidently wield the pendulum and rod for ourselves. It is a fascinating journey that will lead you into areas of your psyche that you never knew existed; the road is not for everyone, but once you make the decision to open the door and see where the path leads, your world will be transformed.

My Story

Many people, like me, come across dowsing at a young age by reading something about it in a book. They may go as far as making or buying a pendulum and practise asking questions of it. Or they may come across a demonstration by someone using the pendulum in a crystal or new-age store and be delighted to discover that it will work for them.

A few folk may even go as far as making a pair of L-rods from some old coat-hangers and a couple of pen cases and have a shot at finding their mains water supply or septic tank; but it's unusual for people to take things further than that, mainly because they have nobody to train them further and can't really see how to apply it in their everyday lives – unless they happen to be a farmer or builder perhaps, where the ability to quickly identify the locations of pipes, cables or other utilities can prove very useful on a day-to-day basis.

Dowsing Magic

I came to dowsing in my late teens, mainly through two books on the subject. One was Guy Underwood's '*The Pattern of the Past'*; a groundbreaking dowsing manual that was much undervalued at the time because it was largely overshadowed by other books in the newly-burgeoning 'earth mysteries' scene, such as John Michell's *'The View Over Atlantis',* or Janet & Colin Bords' *'Mysterious Britain'.* But the one that really inspired me was Tom Grave's *'Needles of Stone'.* Earth acupuncture? Feng-Shui? Megalithic monuments as a form of weather control? Dowsers as modern-day magicians, staking the black streams that brought ill-health? The return of paganism to the land? This was heady stuff indeed, and dovetailed perfectly into my world-view of the time. I had developed an interest in all things 'new-age' and alternative, including a love of visiting stone rings[1] and megalithic remains that remains with me to this day, and a curiosity about our ancient forebears on this land. So one day, a perfectly serviceable pair of wire coat hangers were liberated from the wardrobe and mutilated with wire-cutters and pliers into a pair of L-rods, with a couple of Biro cases for swivel handles; and I set off on my bicycle to see if I could do anything with this dowsing stuff.

The site I chose to go and practise at was one I had been meaning to visit for some time – Bar Hill Roman fort on the Antonine Wall near Croy. This was a key node spot in a network of alignments in Scotland that had been written about by local researcher Harry Bell in a little-known booklet called *'Forgotten*

Footsteps' in 1974. Harry speculated that it was a good candidate for being the lost Druid sanctuary of Medionemeton, which was supposed to be midway between the Forth and the Clyde. It is certainly an evocative location, with the remains of the fort nestling atop a tree-covered hill, next to a pronounced pimple of a crag rising out of lovely deciduous woodland, with good views to the north towards the Campsie Hills. Here I discovered a well in the fort that was near the hilltop, with the water level only a foot or two from the top of the well. With my L-rods outstretched, I walked slowly around the well, and was electrified when they turned in my hands, seemingly without volition on my part. I moved sideways a foot or two and tried again, and once more the rods crossed. Pretty soon I found I was tracing a line that seemed to lead towards the summit of the hill. With rising excitement, I followed this over the hilltop and down the other side, until I found that the line led to what turned out to be the remains of a Roman bathhouse. That seemed pretty conclusive proof to me that I had found a water course, and I headed home, delighted with my new-found skill. Soon I was dowsing pipes around the house, electric cables, coins hidden under a carpet, pretty much anything I could think of. But I hadn't found how to dowse ley lines. That came much later.

Pendulum dowsing also seemed to come easily, and I found I could get good consistent answers to many questions. Yet I always showed respect for the pendulum, and would not employ it for any mundane task, unlike some people who will not make a decision without consulting their pendulum. For guidance I usually turned to the Tarot (another self-taught art) or the *I Ching*. I would often utilise the pendulum in combination with these, using it to pick significant Tarot cards or asking additional questions to help interpret a complex hexagram in *the I Ching*.

At the time, the *I Ching* was becoming very popular in the West, as was the field of *feng-shui*. I studied and trained myself in *feng-shui* as best I could from the scant literature available at the time. I even made my own *lo-p'an* compass with Westernised markings. Nowadays of course, the shelves in bookshops are practically groaning under the number of *feng-shui* books available.

I do inevitably incorporate many of the fundamental principles of *feng-shui* in my geomancy work today, but personally I find that some aspects of it are so culture-dependent that they don't really resonate that well with Western society and so have limited value. For instance, I have trouble relating to the Chinese system of five elements because I'm more used to the Western system of four. But perhaps that's just me, because I'm trying to fit it into an existing Occidental worldview. Besides, there was plenty of home-grown knowledge to absorb from our own cultural traditions – why do people always feel that we need to look to Oriental, Native American, or other aboriginal cultures for their spiritual advancement? I've always been more attracted to

Western disciplines and spent more time researching the likes of traditional witchcraft and ceremonial magic. I read profusely in all areas of esotericism; the Golden Dawn, the Cabala, the works of Aleister Crowley, Israel Regardie, J. H. Brennan, Dion Fortune, A. E Waite and the like, developing what can probably best be described as a neo-pagan shamanic worldview.

My love of visiting stone circles and ancient sites has never left me, but with regards to the dowsing; life and a developing career working backstage in theatre lighting left me little time to develop those skills. I didn't know any other dowsers and had never heard of the British Society of Dowsers, so for several years my dowsing was pretty much restricted to occasional bits of pendulum work.

That all changed in 1999 when I attended an Oak Dragon camp in Cornwall, timed to coincide with the total solar eclipse in August. The camp had a theme of dowsing and geomancy and was being facilitated by Sig Lonegren, who I'd met in Glastonbury a couple of times when visiting a friend, and his geomancy partner Patrick MacManaway, who I hadn't met. I was at that stage of life where I was looking beyond the career and wanting to do something with all the esoteric knowledge I'd been accumulating. I was also desperately keen to witness a solar eclipse, and this camp seemed like the ideal opportunity to forge new life pathways. It was here that I was introduced to earth energy dowsing by Billy Gawn, a remarkable 'deviceless' dowser with a very infectious enthusiasm for the subject. To my surprise, it transpired that I was rather good at it, and I eagerly devoured all the teachings on offer at the camp, which included other aspects of geomancy – astronomy, astrology, sacred geometry, ceremony and so forth. I was already familiar with most of these subjects through my esoteric researches, but I had never really considered that they might be an integral part of geomancy before.

I learned that Sig and Patrick were running a one-year training course on Modern European Geomancy in Glastonbury the following year and determined that I would enrol for this. It meant several weekends driving down from Scotland to Glastonbury over the next year and a half, but it was well worthwhile. Following that course, many of the students continued training and working together in the form of The Geomancy Group. I also joined the Labyrinth Society and The British Society of Dowsers at that time, and nine years later I found myself being elected as President of the BSD. Dowsing really does change your life!

This little book started life as a six-part series of articles for beginners called 'Adventures in Dowsing' that I wrote for the Society's journal, *Dowsing Today*. It proved so popular that the Society would reprint and send them out as background reading for students on their Foundation Courses in Dowsing. Now, after a somewhat longer gap than anticipated, here they are collected and expanded into this volume. I hope you find it a useful guide on your travels with your rod as you explore the world of dowsing. To develop your dowsing farther, I can highly recommend the BSD training curriculum. They have a comprehensive series of courses that will take you from complete beginner up to practising professional, with opportunities to apprentice to existing professional dowsers in the UK. There is nothing that teaches you faster than working in the company of other dowsers. No book or online training course is going to do it, no matter how good it is – you need to get out in the field in the company of other dowsers and learn all you can from them. The BSD has a network of Affiliated Local Groups throughout the country that should be your first port of call if there is one near you. The BSD and their Special Interest Groups also organise several events and an annual Conference. These events are a great way to meet other like-minded folk and try out your dowsing skills, and you don't need to be a member to attend. But you should join the BSD anyway – it's the best way to keep up to date with the world of dowsing, and they produce a world-class journal, *Dowsing Today*.

Find out more about the BSD courses, events, resources and Groups on their website: www.britishdowsers.org.

Chapter 1
Dowsing Theory

So how does dowsing work? Sadly, there is no scientifically accepted theory at present; this is mainly because there is no general agreement in the scientific community that dowsing works at all; mainly because it is not something that fits comfortably into the reductionist paradigm of Newtonian science that still governs academia. Yet anyone who has tried dowsing will know that there is definitely something going on. It is an ethereal and sometimes fickle phenomenon, but with practice results become more reliable and confidence quickly grows. Many scientists have researched dowsing and proposed theories on its mechanism, and in Russia it is still taught as a scientific discipline to students studying for geology qualifications. However in the rest of the western world there remains a small but remarkably vociferous bunch of dedicated sceptics who refuse to even admit the possibility that dowsing could be a very real ability and take every opportunity to publicly vilify anyone brave enough to stick their head over the parapet and speak out in favour of dowsing. Some pretty rigorous testing on dowsing has been done over the years, but results tend to vary depending on who organised the tests and how they interpret the data. Many small but statistically significant successes are dismissed as 'statistical anomalies' by the sceptics.

So regrettably at the moment we are still waiting on a definitive methodology for dowsing to emerge from the scientific establishment, and although there are a goodly number of utility companies and professional bodies who use dowsing in some way, they are usually reluctant to publicise the fact for fear of ridicule. Yet it is not that uncommon to see a water board engineer pull out a pair of dowsing rods from the back of the van when faced with trying to find a leaking pipe or a phone or electric company engineer using rods to track down a buried cable.

This doesn't help us gain an understanding of the dowsing mechanism, and ultimately we each have to develop our own 'map' of the territory; so let's take a look at some possible models. Luckily, the newer scientific paradigms that are emerging today, such as quantum mechanics, string theory, torsion field theory and concepts like the 'information field' can help us with this task.

As we progress, you will see that many aspects of dowsing have a dualistic or binary character, and the main uses of dowsing can be divided into two aspects; **location** and **divination**. Or to put it another way, *finding things, and finding things out.*

When we're looking for underground water, theory and empirical evidence suggests that there is some sort of influence, possibly piezo-electric in origin, given off by water flowing in geological faults and fissures in rock that produces small distortions in the geomagnetic field, and that trained dowsers are sensitive enough to be able to detect these irregularities at the surface. This 'influence' or 'radiation' rises vertically from the water in a narrow band, often passing through several storeys of a structure.

Some research using devices such as scintillometers (a very sensitive type of Geiger counter) and electroscopes has verified the presence of this radiation. It used to be thought that the wood of the traditional 'forked twig' dowsing rod somehow vibrated sympathetically to these emanations, and many dowsers favoured particular types of wood for different tasks; but today it is clear that it is the human organism that is detecting this at a subconscious level, and that the dowsing tool is responding to minute muscle twitches in the hands to give the response – either the twig lowers or raises, the L-rods cross, the pendulum swings, and so forth. There is a common misconception amongst non-dowsers that the rod somehow moves of its own volition, but of course this is not possible. This was written about even as long ago as the 16th Century, and in 1926 Sir William Barrett and Theodore Besterman wrote that, "*dowsing is a purely psychological problem, that all its phenomena find their origin in the dowser's mind, that no physical theory can bear close consideration, and that the movements of the rod and of the dowser have no more direct relation to the discovery of, say, water than as giving physical and visible expression to a mental and abstract cognition*[2]."

As they so sensibly pointed out, numerous experiments have shown that the dowser can detect without a rod in his hand, but there have been no instances of the rod detecting anything without having a dowser attached!

You will often hear sceptics say that dowsing can be dismissed as nothing more than the 'ideomotor response' – the reflex reaction of a muscle in response to a subconscious stimulus - and of course this is exactly what it is.[3] However, the idea of 'earth radiation' to account for the dowsing signal can't explain the divinatory side of dowsing where the dowser seeks answers from the pendulum, nor can it explain how it is possible to dowse for things on a plan or map of the area without being physically present on site. This seemingly magical ability is the biggest stumbling-block for rational minded folk to take on board, and even some tangible target dowsers refuse to accept that it is possible to map-dowse from a distance.

To explain how map dowsing works we are going to have to accept that dowsing is, at least partly, a phenomenon that is interdependent with consciousness. What I mean by that statement is that, at some level, the subconscious mind is privy to information from the physical world that is normally unavailable to the conscious mind, and this information can be communicated to us through the movement of the dowsing tool. So dowsing is in effect a form of controlled clairsentience. Barrett and Besterman came to something of a similar conclusion in 1926:

"*The dowser, in our opinion, is a person endowed with a subconscious supernormal cognitive faculty, which, its nature being unknown, we call, after Professor Richet, cryptesthesia. By means of this cryptesthesia knowledge of whatever object is searched for enters the dowser's subconsciousness and is revealed by means of an unconscious muscular reaction, or less often by man obscure nervous sensation or emotion which produced physiological disturbances, or very rarely by means of direct supernormal cognition made conscious by a visualisation or hallucination.[4]*"

Models

The human brain processes something like 400 billion pieces of sensory information every second; but we are only consciously aware of about 2000 of those. Our conscious mind selectively edits the information to process what we perceive as 'reality'. Psychologists call this filtering process 'latent inhibition', and it is a survival trait that enables us to disregard stimuli that have had no impact on us in the past[5]. Signals that are not important to our survival from moment to moment are simply not processed on a conscious level. This does not mean that the other signals are discarded – everything is processed on some level, but our conscious mind decides what we get to see. We literally create our own reality from second to second. However, our subconscious mind is still aware of the other signals going on around us, and at a deeper level may be able to access the collective unconscious and gain information about other places and events. Many successful 'remote viewing' experiments demonstrate that it *is* possible to access information about distant places, so why shouldn't we be able to dowse about them as well?

This reservoir of external information used to be called the 'Akashic Records' in the language of traditional occultism, but today we are more likely to talk about the 'information field' - a concept based on independent theories by physicist Susan Pockett and geneticist Johnjoe McFadden which suggest that consciousness is created within an electromagnetic field generated by the brain; therefore it is capable of transmitting its informational content to the outside world. Pockett goes a step further than McFadden by claiming that this EM field comprises a universal consciousness that experiences the sensations, perceptions, thoughts and emotions of every conscious being everywhere[6].

If you've done any reading at all on quantum physics, then you may be familiar with the phenomenon known as 'quantum entanglement', which shows that two particles that have been in contact with each other maintain some sort of connection and can transfer information between them, no matter how far apart they are moved. So if we take two paired electrons, then move one to the other side of the galaxy, it will still react *instantaneously* to changes we make to the first electron, in seeming violation of Einstein's Special Relativity Theory. Einstein referred to this as 'spooky action at a distance' and refused to believe it, yet it is now widely accepted and tomorrow's quantum computers are being designed around this principle. The implications of this are pretty staggering – if the 'Big Bang' model of creation is correct and everything in the Universe was expelled from a Singularity, then quite literally, *everything must be connected* at a quantum level.

Another way of looking at it is to think of the Universe as a hologram. If a hologram is broken, each small piece of it contains all the information needed to reconstruct the whole image, albeit at a somewhat lower resolution. So even if we only have a piece of the puzzle to work with, we can extrapolate what the larger picture looks like. This is actually quite a good analogy for map dowsing – often map dowsing produces a picture that is not completely accurate. Usually everything is there, but not necessarily in exactly the same place or at the same scale as the map dowse revealed. I've had this demonstrated to me in my own work on a number of occasions.

Whatever model you choose to adopt, the central concept is that there is a Universal information field that our minds can somehow tap into, and the dowsing tool provides a means of accessing this normally hidden information in our subconscious. By acting as a bridge between our conscious and subconscious minds, the dowsing tool provides a safe and convenient means for us to dialogue with our subconscious; something that is normally only available to people in deep trance states such as hypnosis. There is a well-known self-hypnosis exercise that allows you, once hypnotised, to ask questions of your subconscious self by suggesting that one index finger twitches for 'yes', and the one on the other hand twitch for 'no'. This is essentially identical to a dowsing response, with the exception being that you are not in a trance when dowsing.

Unlike hypnotic or meditational states, brain scans of dowsers at work show that both hemispheres of the brain are balanced and producing brainwaves across all frequencies, so dowsers are not in a trance; they are fully conscious and yet engaged with the deeper levels of consciousness (Figure 2). This chart was created by Dr. Edith Jurka from readings taken using a 'mind mirror' EEG device that measures activity in the left and right hemispheres simultaneously, and shows how the level of activity changes across the various brain frequencies from the most active beta to the least active delta frequencies[7].

Indeed, the correct state of mind for good dowsing is best described as 'engaged yet unattached'. Your mind has to be focused on what you are dowsing for, yet there must be no desire to achieve a particular result.

'Ignorance and apathy' is another good description – *"I don't know what the answer is, and I don't care what it is, but I do need to know".*

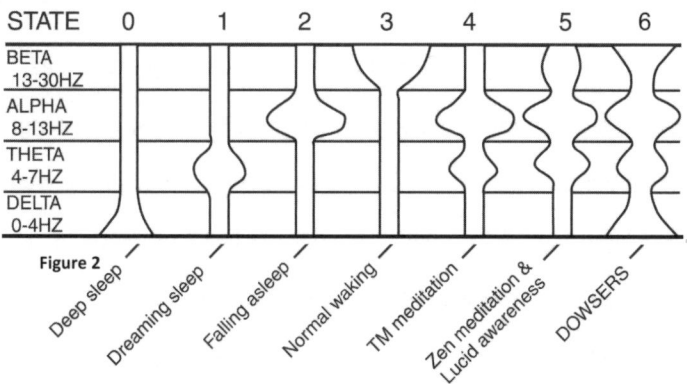

Figure 2

At first, this can seem like being asked not to think about pink rhinoceroses – it's very difficult to think of anything else in such circumstances. So you have to develop ways of distracting the conscious mind from desiring a particular result; in this way you can hear what your intuition is trying to tell you. One way to achieve this is to mentally repeat the object of the search to yourself over and over like a mantra, e.g. "I'm looking for man-made underground structures, man-made underground structures … and so on. Another technique is to try inducing a state of child-like innocence as though you earnestly desire to know the answer but have no idea what it will be and are looking forward to the surprise of finding out – "I wonder what the answer's going to be?" In either case the repetition of the phrase will be enough to distract the conscious mind enough for the dowsing reaction to manifest.

The other essential attribute for accurate dowsing is to ask clear and concise questions about the object of the search. It's no good just asking for "underground water" if you're looking for a place to sink a well or borehole – this could find water in pipes or septic tanks as well as aquifers or water flows. A better question would be "potable water flowing all-year round in underground streams". But even this might not cover all the things that you need to know. What if the water is contaminated? If so, what are the contaminants – salt, iron, sulphur, coliform bacteria, radioactivity? What percentages of contaminants? What is its pH content? There are a lot of questions that you could be asking just to find a good borehole site.

Or, if you're looking for a lost pet, it is better to say "Where is so-and-so's pet collie Fido *at this moment in time*", rather than simply "Where's Fido?" Time is always a factor in divinatory dowsing, as we can focus our minds to dowse about any period of time in the past, and even to a lesser extent on the likelihood of some future event coming to pass. It is always important to factor in exactly what time period you are asking about.

You can see from this how vital it is to be able to clearly frame a series of pertinent questions for each dowsing task. The more knowledge you have about your subject, the more precisely you will have to frame your questions. Indeed, it is advisable to write out a list of possible questions that you think you might need to ask before starting your dowsing session. You will probably have to alter them as you go, but at least you have a framework to begin work from, and the process of deciding what questions you want to ask helps to clarify your thinking and get you into the 'dowsing zone'

It is important to realise that there is no inherent magic in the dowsing tool. This is a common misconception by non-dowsers and beginners, and one that is propagated by new-age shopkeepers who have extensive (and expensive) stocks of crystal pendulums to sell. Don't be fooled into buying an expensive pendulum because you've been told that it's stronger or somehow better than your home-made nut-and-string combo. Of course, you can buy the expensive pendulum because it looks pretty and you like it, but that won't make it work any better. As long as the pendulum has sufficient weight to it to give a nice swing that you are comfortable with, it can be made of whatever material you like. The reaction happens because the muscles in the hand twitch very subtly in response to your subconscious stimulus, and this small movement is amplified by the dowsing tool. The tool itself is irrelevant – it's the muscle response we're interested in. All we have to do is learn how to control and interpret this reaction.

Dowsing Magic

As we progress with our lessons, you will find that this process becomes easier and easier. As you gain confidence, you become more relaxed with the reaction and your muscular movements become larger and easier to detect. You often find that advanced dowsers are criticised for 'cheating' because you can see their wrists move, but this is perfectly natural and just means that they have learned to relax their muscles enough that they movement is visible. Eventually it becomes so automatic that you often don't need to use a dowsing tool, and at this stage you are a 'deviceless dowser' and can operate without tools, which is the ultimate goal for many of us.

Chapter 2
Pendulum Practice

The simplest dowsing tool is the pendulum, and it's what nearly everyone starts to learn with. If you haven't got a pendulum to hand, you can easily make one with a small weight and a piece of fine cord or strong thread. A steel hex nut or a lead fishing weight is ideal. A bunch of keys tied to a short piece of string will do if you can't find anything else. The cord should be as thin as you can manage; a fine chain is also good. Try not to use a pendant hanging on a loop of chain as the double length may affect the swing of the pendulum.

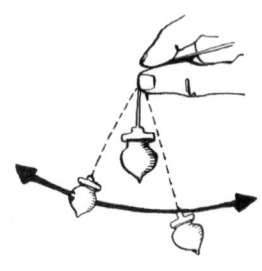

Figure 3

Hold the cord or chain between your thumb and index finger of your dominant hand with the finger pointing downwards. The idea is to minimise the contact area between the cord and your fingers. Keep your arm and shoulder relaxed but free to move. Don't rest it on your knee, the arm of your chair or anything else as it may adversely affect the swing. Experiment to find what length of cord gives you a moderately rapid swing – usually around three or four inches of cord is ideal (Figure 3). You don't want the cord too long as it will make your responses sluggish and ponderous; similarly you don't want it too short as it will make your responses too unpredictable. Swing the pendulum back and forth until you are comfortable with the feel of it and the length of the cord or chain.

Training our responses

Many people, when they first start using a pendulum, will sit with it stationary and wait for it to move in response to the question. Although this approach will work, it becomes very laborious if you are dowsing through a series of questions, as each time the pendulum has to overcome the inertia of being at rest and it takes time to get up to speed. A far faster approach is to start with the pendulum swinging back and forwards. We call this our 'search' or 'neutral' position. When the pendulum is already moving this way it's much quicker to give a response.

Next, we have to establish our code for communicating. This will take some time to develop fully as many reactions are possible, but for the moment we're only interested in our 'yes' and 'no' responses.

Dowsing Magic

Sit comfortably with your feet apart, and start the pendulum swinging back and forth between your legs. Each time you do this say to yourself, "This is my search position". When the pendulum is swinging comfortably, ask it to "show me my 'yes' answer". Most people will find the pendulum develops a clockwise swing. You can help reinforce this idea by holding it over your right (dominant if you're right-handed) knee. This develops the idea of the body having a polarity – positive on the dominant side, negative on the non-dominant side (note positive and negative are only terms for comparison here; don't start thinking of them as 'good' and 'bad' in any way).

Once you're happy that you're getting a good reaction, repeat the exercise by asking for your 'no' response. Again, you can hold the pendulum over your non-dominant knee to reinforce the idea if you like. Most people here find that they get an anti-clockwise swing developing, but it's possible that you get something else, such as a side-to-side swing. Once you have programmed these reactions, they will always be the same for you, no matter what pendulum you're using (I once found one of my students was asking for her basic responses with each new pendulum she used!). Don't change these responses unless you have good reason to, otherwise you will get to a point someday when you cannot remember what your yes response is, and will never feel confident enough to trust your dowsing.

If you're having trouble getting any reaction at all, try sitting with your eyes closed while the pendulum is swinging in the search position. Now visualise the swinging pendulum in your mind, and mentally ask the "show me my 'yes' response" question, whilst picturing the pendulum moving into the desired pattern. Then open your eyes and look – the pendulum should be doing what you want it to do.

In really difficult cases, it is possible to actively program yourself by forcing the pendulum swing into the pattern you want, whilst affirming to yourself "This **is** my 'yes' response" and "this **is** my 'no' response". Sometimes it takes a while for the brain to get the message. You will need to re-affirm these responses several times until you are happy that you are not forcing the response any more. This is not 'cheating'; remember, all you are doing is programming your brain to accept the dowsing reaction.

Once you are reasonably confident with your responses, devise a list of twenty questions that you can dowse for. Try to ask questions with verifiable true or false answers, such as "is my wife's name Sue?", "Is my name John?" "Do I live in London?" and so forth. Practise by asking yourself these kinds of questions until you are confident with the responses that you are getting from your pendulum.

Try also the traditional 'three-card Monte' dowsing trick of identifying one red card out of three face-down cards, or have someone place a coin under one card and try to find that; try to identify a pound coin hidden under a tablecloth with a bunch of other coins; try to identify if there is current flowing in an electric cable. There are loads of little tests like this that you can devise for yourself and practising at these is the best way to develop your dowsing before you get into areas where it's not always possible to get a verifiable answer at the time (for example dowsing for a borehole – you won't get confirmation until the hole is actually drilled).

Here's a simple questionnaire that you can use to find out if you have any food intolerances (

Figure 4). This could apply equally well to any meal situation, or indeed any food; it doesn't have to be 'last night's dinner' as referred to in the form. Fill out the initial fields, and then use your pendulum to answer the three questions for each one:

Exercise – what did you have for dinner last night? *List main ingredients; e.g salmon, potatoes, carrots, peas, tartare sauce etc.*	Did this item fulfil my nutritional needs?	Is my system intolerant of this item?	Is this item detrimental to my health?

Figure 4

You can experiment with the questions asked - for instance, there is a subtle but distinct difference between the second and third questions, yet they could be seen as meaning the same thing. Can you think of better questions to ask?

If you are asking personal questions about somebody else make sure you have their permission before dowsing and always keep your dowsing ethical. It is both impolite and unethical to use your dowsing to enquire into somebody's private life without asking their permission. A good rule-of-thumb is not to use dowsing to ask anything that you wouldn't ask them to their face.

In many books on dowsing, you will see sectored pendulum charts for dowsing, where different areas of the chart provide different answers. You hold the pendulum over the centre point, and the angle of its swing determines which sector of the chart contains your answer.

Although these are popular with publishers because they look pretty and fill up the page, I personally don't believe that they are a good tool to use with pendulum dowsing, especially when you are just starting out. Having a multiplicity of possible answers dependent on the angle of the pendulum swing will only confuse your normal responses. It is far better to devise a list of possible questions before you start dowsing and keeping to your yes/no responses as you work through the list. Later on, when you have developed confidence in your dowsing, you can try experimenting with these sectored charts to see if they work for you.

As you get more confident with your pendulum, you may find other reactions developing. For instance, if I get a weak oval swing in a clockwise direction, it means 'yes-maybe' (and conversely for 'no-maybe'). In this case I'll try and refine my questioning further.

I also get a side-to-side motion, which for me interprets as 'impossible to answer - the question doesn't make sense'. If my pendulum simply continues in the 'search' position, for me it means 'don't know'. In either case, I know I have to ask a totally different question to try and come at the issue from another angle. But such responses will develop naturally for you over time. If you find your pendulum doing something unexpected that you haven't seen before, you will have to come up with a series of questions to ask to work out what the new reaction means for you.

Pre-flight checklist

Ok, if you've got this far, you're almost ready to start finding stuff in the field; but there is one more thing we have to cover first, and that is the issue of ethics and protection. Some dowsers disdain this as unnecessary, but I always take the attitude that it's better to have protection than not. This is especially true if you're doing dowsing at an ancient site where you may run into some strong energies. You wouldn't go out on a wintery day without putting a coat on, and likewise you shouldn't go out dowsing without taking some simple precautions.

Fortunately, dowsing has built-in safeguards that can and should be activated every time you try it. Here's a quick test you should do before you start any dowsing session, or at each new site.

With the pendulum swinging in your 'search' position:

1. State what you want to do (e.g. "I want to dowse this stone circle for water lines", or "I want to find my sunglasses" and so on). "Is this appropriate at this time?"

The pendulum should start to show a 'yes' response if this is OK.

2. Ask "CAN I do this?" (Do I have the dowsing ability for this task?)

3. Ask "MAY I do this?" (Do I have permission from the site guardian/ landowner/ Spirit of Place to do this?) It always pays to check in with the site guardians before entering stone circles or other sacred space. The last time I visited Wayland's Smithy, I asked my questions as I walked up the access path from the Ridgeway. Just as I asked this question, a huge fluffy bumble-bee came right up in front of my nose, hovered for a moment, then turned and flew ahead of me down the path. I took that as a 'yes' answer!

4. Ask "SHOULD I do this?" (Is it safe for me to do this? Is it appropriate for me to do this now? Is there anything I've forgotten? Are there any unforeseen factors that I don't know about?). I once got a 'no' for this question

when I wanted to dowse a neglected and overgrown stone circle in Perthshire. It was then I became aware of a nest of newly-hatched chicks behind one of the stones; clearly not a good time for me to be tramping about in the circle.

If you get a 'no' response to any of these questions, postpone your dowsing to another time. You can try rephrasing the first affirmation a bit to narrow it down, but if you're not confident of any of the answers then come back and try again another day. But if everything checks out, then go ahead, ask your question (e.g." I want to find any water veins in this area") and dowse. What you are looking for is only limited by the questions you can formulate in your head. And the more precise these are the clearer the results you get. It's sometimes difficult to phrase your intent in a yes/no question, but persevere; try a different angle on it and narrow it down gradually. Remember that it helps to write out a list of questions before you start dowsing.

The above protocol is one of the simplest techniques to get your mind into the 'dowsing zone' and ensuring that you are not going to be hit with anything unexpected. Take a little time to get relaxed and focused before you start dowsing, especially if you've just driven several miles to visit a special site. Allow your body 10 or 15 minutes to relax and engage with the site before you start dowsing. Intensive dowsing can be very draining, so don't try anything major if you have a cold, headache, or are otherwise feeling run down. You'll only make yourself feel worse.

Dowsing Magic

As your dowsing skills progress and you learn to manipulate energies in addition to merely detecting them, you will probably want to modify the basic protocol with additional steps to make sure that you are properly and completely attuned to the environment and to the task at hand; for example clearing yourself of any non-beneficial energies and making sure you are properly grounded. But the 'Can I? May I? Should I?' routine is fine for most general dowsing situations.

Chapter 3
Historical Context

Since time immemorial, folk have been using dowsing to locate things, particularly water. We don't know much about the early history of the art, but it seems to be indigenous to several different cultures. Quite possibly the earliest recorded reference to dowsing is in the Old Testament, where Moses is told to 'follow his staff' to find water by God, and strikes a rock with the staff, causing water to flow.

The first *printed* references to dowsing show up in the 16th Century. Georgius Agricola's 1556 book 'De Re Metallica' (*On Metals*) discusses the art of dowsing to locate underground metals and minerals, and contains many woodcut illustrations of dowsers at work (Figure 5). In the illustration, you can see a dowser cutting a suitable branch from a tree at the top of the picture, and two examples of dowsers at work (marked 'A'). This book became the standard reference text on mining

Figure 5

for nearly 200 years. Surprisingly, the book did not appear in English until 1912, when it was translated by one Herbert Clark Hoover, who later became the thirty-first President of the United States.

In 1561, Queen Elizabeth I needed some ready funds to clear an inherited deficit and build more ships for the English navy, heavily engaged against the Spanish Armada. The existing silver coin of the realm was badly debased through the widespread practice of 'clipping' – shaving small slivers from the edges of coins and melting them down – and so she planned to issue new coins with the silver content reduced by mixing it with 15% copper, thus effectively skimming 15% of the value for her own coffers. But first she needed to find an abundant source of copper. She asked her chief advisor Sir William Cecil to engage the services of one Marcus Steinberger, a mining expert and dowser from Augsburg, to explore for copper around Coniston in the Lake District. The museum in Coniston has a small display about the copper mining, although sadly it doesn't mention the dowsing connection. Another German expert, Christopher Schutz, was later brought over and discovered zinc ore in the Mendips. This may have been the key moment in introducing dowsing to English miners, although there are said to be earlier stories of Cornish miners dowsing for tin[8].

Nowadays dowsing as an aid to mining has largely been superseded by more scientific prospecting methods, yet dowsers are still sometimes employed to map-dowse for oil and gas deposits[9], and dowsing for buried treasure is still a popular activity with some, even in this age of metal detectors.

There has always been a degree of mysticism associated with dowsing, such as beliefs that specific woods had affinities for different metals, that the rod was moved by divine influence (or diabolic if the desired target wasn't found), and so on. Even in the late 19th century an old Hungarian widow was observed to hang the corpse of a black hen on a hazel tree from which she then cut a dowsing rod at a single stroke while uttering a prayer. The rod was then taken home, stripped before sunrise on New Year's Day, and moistened with blood from her left arm and leg. During this procedure the woman intoned: *"I give blood to the rod. She gives me its bark which I consume; in this way we are bound to each other, betrothed as it were, and when the moment is come when my strength has reached the time when it must break out, then the rod must show me exactly where is hidden the treasure which I am appointed to raise."*[10]

Thankfully things have moved on a bit since then – can you imagine going through that routine every time you wanted to do a bit of dowsing! Treasure hunting has always been a popular dowsing activity, and many of the superstitions attached to dowsing relate to this. But before you get your pendulum out and start dowsing for lottery numbers, remember that dowsing

Dowsing Magic

works best for *need*, not greed. It's very hard to maintain an air of objective neutrality in circumstances where personal gain is involved.

Figure 6

One of the best recorded dowsing tests took place in 1913 in Paris. A group of dowsers led by Henri Mager (Figure 6) successfully detected a series of underground caves and mineshafts that were otherwise recorded only on a series of unpublished 1:10,000 charts known only to a few people on the city Council. The dowsers even identified areas where 'pillars' of rock had been left unhewn to hold up the roof of the mine.

Mager is best remembered today for his invention of the 'Mager Rosette' (Figure 7), a disk with coloured segments on it used as a 'witness' – a device to help focus the mind on the object of the search in dowsing. Mager also hypothesised that currents of a physical force similar to electricity moved from underground water up through the dowser and into the rod, passing in a screw-like fashion up one of the branches of the forked

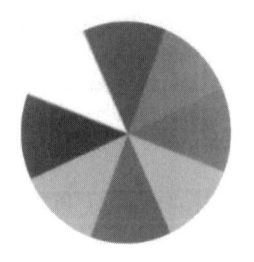

Figure 7

stick and returning down through the other, back through the dowser into the earth, and finally back to the source of the underground water, thus completing a circuit. He further believed that having leather or rubber soles on his shoes or boots would prevent this force from coming up through him. Any modern dowser who has managed to dowse whilst wearing Wellington boots will know that this is simply not true, and is an example of what some dowsers call a 'shibboleth' – an outdated concept that acts as a self-imposed mental limitation. The only constraints put upon your dowsing are those that you put there yourself.

In 1929 a German aristocrat and eminent dowser, Baron Gustav Von Pohl, made the extraordinary claim that cancer was caused by 'energy currents' that ran under the beds of the people that had died from the disease. To test his theory he persuaded the Mayor of the Bavarian town of Vilsbiburg to allow him to survey the town for such energies with his dowsing rod. Von Pohl's dowsing map was then compared with the town's record of houses where people had died from cancer during the last ten years, and showed a remarkable coincidence. Precise examinations were next made of houses only partially affected by the water veins and the beds of cancer patients were found to stand directly above them.

Cancer specialists rejected his report, claiming that the town had such a high incidence of cancer that his conclusions meant nothing. Undeterred, von

Pohl asked the Bavarian government if he could repeat the experiment in Grafenau, the town with the lowest incidence of cancer. His survey had exactly the same result – all the cancer victims in the town were found to have been sleeping over water veins.

The Baron was then asked by medical officials of other towns to make similar measurements, which gave like results. When measurements were taken in the City of Stetten, Dr. Harger, chairman of the city's medical scientific association, declared that all such 'deadly earth currents' ran beneath the beds of all the 5,348 people who had died from cancer during the last 21 years. This inspired some scientists and medical engineers to probe into the matter a little more deeply.

A few years later, in the 1930's, French researcher Pierre Cody placed gold leaf electroscopes in cellars of houses of cancer victims. Devices placed over the water veins detected an ion concentration 10 times higher than the control devices placed over 2 metres away. Repeating the experiment at closer and closer distances between the electroscopes produced the same result, until the electroscopes were only a few centimetres apart; revealing that the radiation rose in a very tight vertical band from the underground water and did not diffuse laterally.

Lead sheets placed under the electroscope retarded the discharge of the electroscope, and after a month, displayed discolouration matching the radiation zones; implying that the radiation is actually ionising radiation – i.e. is slightly radioactive.

Baron von Pohl's theory was re-examined in the 1950's by German scientist Dr. Joseph Wüst, and Jakob Stängle, an engineer and dowser. Dr Wüst conducted experiments in several German towns in the rooms of cancer patients and found small increases above background levels of radiation in the houses he surveyed using a Geiger counter.

Stängle had a very good reputation of locating water-bearing fissures in crystalline rock, using a device he had developed over a period of 15 years called a scintillation counter, which was much more sensitive to gamma radiation and other charged particles than a standard Geiger counter. As it is wheeled along, it registers variations in the background radiation, and it was particularly effective at detecting water lines and oil deposits.

He first carried out a survey in the French town of Moulins where a local physician had claimed that cancer victims in the town dwelt above geological faults. Using his scintillation counter, Stängle again found unusual peaks above normal background radioactivity over these faults. Later, Stängle visited Vilsbiburg where he was able to confirm Von Pohl's earlier findings.

Another contemporary researcher, Dr. Armin Bickel was an astrophysicist who studied in Germany with Werner von Braun and other rocket scientists at Peenemünde, developing the V2 rocket. In common with many other German rocket scientists, he migrated to America and worked for the fledgling US space program after the war, and eventually found himself challenged by NASA to build devices that could be mounted in satellites to chart the earth's mineral deposits from orbit.

After retiring, he used the same technology to build a super-sensitive scintillometer he called an isotope detector that was much smaller than Stängle's device and was able to be slung around the neck instead of being wheeled along on a trolley. It works by sensing weak gamma emanations from anomalies in the earth. When there is a deposit of ore under the earth directly below the sensor in his machine, the sensor will register a very weak signal that identifies the element by its characteristic gamma energy level. With this device, it is possible to detect oil and mineral deposits from a great distance, even from an airplane or satellite.

Interestingly, Bickel was also a dowser, and would frequently use his L-rods to check the machine's results when it appeared to be malfunctioning![11]

Subsequent studies of the geomagnetic field over water-bearing fissures have found that there are slight variations of a few thousand nanoTesla in these areas.

These variations in radioactivity and geomagnetism are extremely subtle and difficult to detect even with expensive scientific equipment, yet they are relatively easily found by dowsing. Our current dowsing model of geopathic stress stems originally from these pioneering studies.

In more recent times, the American marines were taught dowsing to help locate underground tunnels in Vietnam. The marines applied themselves to this skill with military precision, and were soon using it to locate not only tunnels, but also weapons caches, pits, landmines, tripwires and other booby-traps. They even discovered the secret communications system of the Vietcong, who would hide messages in bamboo tubes buried under the jungle trails[12].

In 1915, during the First World War, Australian troops at Gallipoli found themselves without any provision for the supply of water, because no-one at General Headquarters thought that the fighting would be very prolonged. In just over 24 hours, Lieutenant William Dawkins of the 2nd Field Company of Engineers had dowsed the locations for 20 shallow wells, which provided a supply of 20,000 gallons a day of good drinking water, more than sufficient for the troops and horses[13].

The British Army has also used dowsing to locate water supplies for field camps (we shall look at an example of this in the next chapter), and to search for mines and other unexploded ordnance. Indeed, it has been said that dowsing is the most reliable method of mine detection, although to be successful at this requires a high degree of confidence in one's dowsing ability! There has always been a strong military connection with dowsing, and The British Society of Dowsers was founded in 1933 by members of the Royal Engineer Corps.

Chapter 4
Working with Rods

Traditionally the most common dowsing activity, and still the most practical application of dowsing, finding water is a skill that is usually the first thing that the novice dowser learns to detect. Most people will be familiar with the traditional 'forked twig' or V-rod dowsing tool, but there are several other tools available to the dowser. We have already looked at the basic pendulum; in this section we will learn about some of the others.

An article in by Major Richard Neve in 'Pennant', the Forces Pension Society magazine, mentions a Sea Dayak (Iban) tracker from Borneo known as 'Pte. Dadop' who dowsed for water by cutting two pencil-width twigs and laying them on the palm of his hand in a 'V' shape. In the presence of water, the point would raise upwards. The soldiers were encamped on top of a ridge with no ready access to a water source:

> 'While the water bottles were being collected Dadop had walked round the base area looking at various bushes. Spotting what he wanted he cut off two pencil width pieces of branch, each about three inches long. Dadop leading, I gave the order to move off. He immediately placed the two pieces of wood in the palm of his hand so they formed a V with the base pointing towards his fingers. He then set off at a fast pace down the side of the ridge... It suddenly dawned on me I was watching the Iban

equivalent of water divining. After descending no more than 100 feet he veered off into a slight hollow about 30 feet across.

"Dig here, Tuan".

We quickly dug down two feet with an entrenching tool and our parangs. The resulting hole soon filled with slightly muddy but drinkable water deep enough to submerge each water bottle in turn until it was full. In half an hour we had finished and were on our way back"[14].

This is just one example of the myriad tools and methods that have been developed by dowsers over the years. Some, like the late Guy Underwood's 'geodetic rod', have proved too idiosyncratic to be generally popular (although it still has a few dedicated adherents); others, like the angle- or L-rod, have become almost universal favourites. We'll look at more tools later on, but for now we'll concentrate on the two most popular styles of rod – the 'Y'- or V-rod and the L-rods.

The V-rod

The classic 'forked twig' is still the favoured tool amongst water diviners. It has the advantage that it can be cut on-site. But it will also crack or break after a few uses, and although easily replaced, many dowsers will prefer a more sturdy replacement such as plastic 'V-rods'. Two suitably whippy lengths of plastic rod or conduit are held together at one end by tape or sleeving (Figure 8). The length is not critical, but around 40-50cm works well.

To use the V-rod successfully, you need to get the tension correct. Hold it palms up, with thumbs facing outwards and the point facing away from the body. It should feel like the rod is alive and trying to twist out of your hand and you are only just managing to keep it steady.

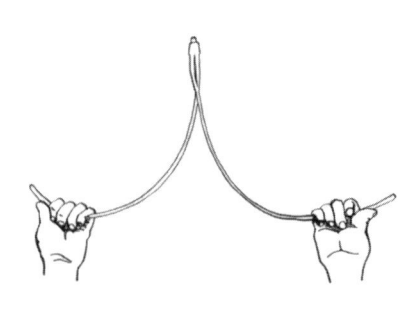

Figure 8

Try to keep your forearms parallel and don't let them get too far apart. If there isn't enough tension, try turning your hands inwards a little, so that your little fingers are moving closer to your body and your thumbs move forward. It's difficult to explain, but unmistakeable once you've got it right.

The V-rod is a very responsive tool, and there is no mistaking the reaction when you find the target; the rod twists downwards or upwards, often violently enough to break a wooden rod. That's also its main disadvantage – it can give you a nasty crack on the nose (or other more painful places) if you forget to hold it away from the body. Sometimes the rod will even spin completely around in the hands. Just like with the pendulum, you can interpret these reactions as 'yes' and 'no' – usually down for 'yes' and up for 'no'. Spinning completely around is usually a sign that you have passed the target spot, but may mean something different to you. As we learned with the pendulum, this is a signal that you will have to interpret for yourself.

If there is a limitation to the V-rod, it is its lack of variety. It is difficult to get anything other than a 'yes' or 'no' answer from it, whereas other tools like L-rods can provide more versatile responses. However, it could be argued that its very lack of ambiguity is the strongest selling point of the V-rod; there is no doubting the response! Yet with a little creative ingenuity, the V-rod can be persuaded into an enhanced repertoire of responses; for example, you can use it to determine direction of flow in a water vein or pipe by turning round in a circle whilst standing on the line; the rod will react when facing 'downstream'.

Perhaps the forked twig owes its long popularity to its natural nature. A suitable branch can be quickly sourced in a hedgerow wherever the dowser is working, and is easily replaceable. But there is also a more sinister historical reason for its popularity. Dowsing was denounced by Martin Luther during the Reformation, and for many years it was a crime to use dowsing for anything other than water finding (this being recognised as a vitally important activity). So being seen carrying a dowsing tool about could lead to a visit from the Inquisition in many places. Far better to just cut a rod on site when you needed to work, and if anyone asked any questions it was a simple matter to snap the rod in two, tuck the pieces under an arm and claim that you were simply collecting firewood.

The L-rod

It's probably fair to say that most dowsers use the L-rod these days (Figure 9). Like many people, I made my first pair from some old wire coat hangers and some Biro cases. Some like lighter rods with swivel sleeves; others prefer a heavy solid L-rod. Swivel rods are probably easier for beginners who may have a tendency to grip the rods too tightly; but there's nothing like feeling a solid rod twisting in your hand

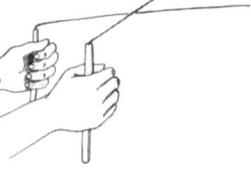

Figure 9

when you hit the target. Heavier rods are more useful in outdoor situations where it is windy, and the heaviest pair I have seen were over a metre long with handles about 50cm, made from solid copper rod. They were so heavy I was unable to hold them steady for more than a few seconds, yet the dowser who had made them swore to their effectiveness in archaeology work.

Hold the rods grasping the short ends in the fingers of your fist with the long ends sticking out in front of you like a pair of pistols. This is the 'search' position for L-rods. Make sure that the long edge isn't resting on your fingers as it needs clear room to move, and keep your fist *relaxed* - the most common mistake that beginners make is to grip the rods too tightly. Point the rods slightly downwards, and then slowly lift the ends up until just before they start to swing outwards and go out of control. This is the most sensitive position for the L-rods.

Rods can react either by crossing over or opening apart. Usually they will cross when 'on target'. You can also get 'yes' and 'no' reactions from L-rods, for most people they cross for 'yes' and swing out for 'no'. Like with the pendulum, this tends to work best when the rods have some momentum, so try asking your questions as you are walking forwards.

Dowsing Practice

We'll look at more tools later on, but by now you should be confident enough to find things in the field. Start by dowsing for utilities like drains or water pipes around your house, or try to find your septic tank if you have one.

You can even try dowsing for a rope or cable that you just lay out on the floor – if nothing else this will increase your confidence in your reactions.

When you have gained confidence with these exercises, move on to underground water streams. Water has a strong reaction signature so it's easy for even novice dowsers to find. Remember to focus your mind on the object of your search but try to remain detached from the desire to find it. Think about streams of water flowing in fissures in rock, imagine the sound that would make, imagine what it would feel like water was running over your bare feet – the more vivid a picture you can conjure in your imagination, the easier it will be to dowse. Some people find that it helps to have a drink of water before starting, to 'program' the brain into looking for that. It's also a good idea to find an area free from human intervention so that you are sure you are not being misled by buried pipes, cables or other man-made artefacts. Somewhere like a large park or a rural green field is ideal to find your first underground water.

When looking for water streams, remember we are looking for thin veins of water flowing in fissures in the rock deep underground. Try to focus on the idea that you are looking for the centre of the vein. When you find this, go back and approach it again, this time telling yourself that you are looking for the nearside edge. Then continue, this time looking for the far edge. You should find that you have identified a spread of lines just over a metre wide in all. Together with the centre line, this 'triad' or 'triplet' reaction is typical. It doesn't necessarily mean that the water vein is exactly that width, but it seems to be the energetic 'signature' of water. Energy leys can present themselves in a similar fashion, but usually the width between edges is around three metres or more. We'll look more at these later.

Once you've identified a water vein, move to one side or the other of your first point and dowse it again so you have an idea of its orientation. You can also ask your rods to show the direction of flow. With L-rods just ask as you approach the line and the rods will swing to one side or the other. You can even do this with one rod, which is actually easier. If you're using a V-rod, stand on the line and turn around on the spot – the rod will react when you're facing 'downstream'.

With L-rods, it's even possible to follow the line of the water; you will find that the tip of the L-rod seems to 'stick' to the water line, weaving gently from side to side as you walk along. However, water veins often have little spiralling side flows that can mislead you, and you may suddenly find yourself walking in rapidly decreasing circles. To accurately track the course of a stream it's usually better to repeatedly criss-cross the ground in a zig-zag pattern and place a marker each time you get a reaction (Figure 10).

Dowsing Magic

The same basic principles apply if you're looking for pipes or other utilities around your house, except of course you will be asking for 'water flowing in pipes'. It's important to get this distinction clear in your mind; once I had dowsed a water vein on which I planned to site a small standing stone, and some months later when I went back to dig the hole for the stone, I couldn't find the water vein.

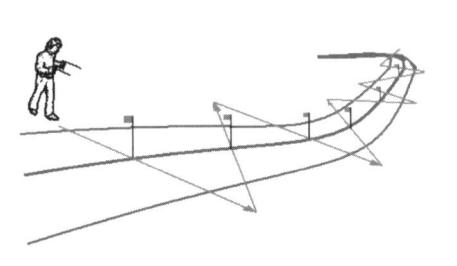

Figure 10

Undeterred, I started digging anyway, only to come across a plastic field drain about a foot under the surface. At the time of my original dowsing it had been November and there had been water in the pipe; when I went back it was July and the weather had been dry for some time. That explained why I couldn't dowse any water on my second visit – the pipe was bone dry! But it was both a good confirmation of my dowsing and a reminder that I had to be more specific in my mental questioning in future. You often find yourself getting little confirmations like that as you are learning, and of course it's a great confidence booster.

The Hydrological Cycle

If we are to be serious about our water dowsing, we need to know a bit more about how water behaves in the ground. A good water dowser will have a sound working knowledge of the geology where he is working.

The generally accepted picture of the hydrological cycle that you may remember from school geography lessons is that water falls from the sky as rain or snow, flows downhill as streams and rivers to the ocean, where it evaporates back into the sky to repeat the cycle. Some may percolate deep underground via caves, fissures and permeable strata like gravel or sand, but ultimately it finds its way to the surface again. Water vapours from power plants, vehicle exhausts and emissions from factories are also a factor in the evaporation cycle, but ultimately there is a fixed quantity of water in the world.

The height of the available water table is dependent on the highest point of rainfall in the surrounding landscape – this is how we can have water pressure in our homes without pumps in most cases. In addition, the pressure of overlying rock

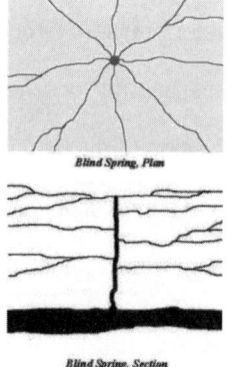

Blind Spring, Plan

Blind Spring, Section
Figure 11

strata will increase the nominal height of the water table by creating a virtual 'pressure surface'. This determines the natural height to which water will flow, and it may be several metres above the actual ground level. Boreholes whose tops emerge beneath this 'pressure surface' will flow naturally, but those whose tops are above the pressure surface will need to be pumped.

Yet this 'artesian' picture does not apply to all geological conditions, and dowsers have proved many times that it is possible to find underground water near the tops of mountains, where this standard picture of the 'pressure surface' cannot possibly apply. So what's going on? An alternative theory proposed over 100 years ago by a Stockholm professor of mineralogy called A. E. Nordenskiöld and later revived in the United States by Bavarian-born mining engineer Stephan Reiss[15], postulated that water can be formed though chemical reactions under great pressure deep in the Earth's crust; or perhaps water that has percolated down far enough is evaporated and forced upwards under pressure through fissures and faults, where it eventually re-condenses. This 'primary' or 'juvenile' water as it is sometimes called is highly energetic in dowsing terms, and is nearly always found at sacred sites in the form of a blind spring, also called a 'dome' by American dowsers (Figure 11). Here, the rising water is unable to break through to the surface due to some impermeable rock strata, and it flows outwards in smaller veins, looking rather like the spokes of a wheel in plan view. Blind springs generate a strong vortex of energy above them, which characteristically produces a helicopter-like spinning of L-rods when dowsing that is quite spectacular. The first time I dowsed a blind spring I had no idea what strange force had suddenly possessed my rods! Fortunately, a more experienced dowser was on hand to explain things.

You'll find a similar but weaker vortex over the crossing point of water veins, and also in some earth energy situations, but more on that later.

The presence of underground water is often signified by natural signs; certain plants like yarrow and hellebore like to grow over water veins, and trees will tend to lean towards a strong water line until their branches are within the vertical band of radiation that rises from it. Figure 12 shows an extreme example of this, taken at Mdina in Malta.

Figure 12

Trees that are situated over blind springs or other vortices will typically develop a strong corkscrew twist to their growth pattern, and in extreme cases where there is a group of trees you may find that there is an explosion of growth in all directions.

Chapter 5
Lost Objects and Missing People

The disappearance of Madeleine McCann in 2007 created a flurry of interest in the use of dowsing to locate missing people. A number of calls were received at the offices of the British Society of Dowsers, some from non-dowsers asking why we weren't using dowsing to locate her, and an even greater number from dowsers claiming to have located her - unfortunately citing a wide variety of disparate locations[16]. This highlights one of the main difficulties arising when a number of dowsers work on such emotive cases - namely the high noise-to-signal ratio. With so many different reports it is inevitable that many are incorrect ('noise'). Discerning which ones are worth following up on ('signal') requires a large input of time and resources; resources which might be better employed pursuing more conventional lines of inquiry. It can also be difficult to get a "fix" on a living animal or person, mainly because they tend to move around. Inanimate objects are much easier to locate. One of my early cases involved map dowsing for a missing dog. Working from an OS map of the area, I got a very strong fix on an area about a mile from the dog's home, and reported it to the owner, who promptly went out in her car to search for the animal, sadly without success. However it later transpired that the dog had been seen by a local bus driver in the place I had dowsed at approximately the time I had dowsed it. Clearly the dog had plans for farther adventures in mind and had gone on exploring! He turned up a few days later, very dirty and near-starving, so the story ended well.

So although it may seem easy to do in principle, it is a rare dowser indeed who gets consistent results from finding people or animals. If everybody got it right every time, we would be living in a very different society – perhaps one similar to that portrayed in the film 'Minority Report', where potential criminals are identified by psychics and apprehended **before** they commit the crime. When the police or other official bodies *do* use dowsers to find missing people, typically the dowser is not informed of the results afterwards so they have no way of knowing if they were right or wrong. In a perverse kind of way, this actually helps the dowser as it reduces any emotional attachment to the results, and the only gauge of success is whether the client uses them again on future cases.

Another factor to consider if you are undertaking this sort of work is the risk that your dowsing might be **too** accurate, and you suddenly find yourself a suspect in the investigation. I have heard that this has actually happened to a remote viewing group in the USA, who suddenly found the FBI knocking on their door when one of their sessions unearthed some highly sensitive material. In today's paranoid global climate, there is a very real danger that you might find yourself arrested under the Terrorism Act if your dowsing strays

Dowsing Magic

into such restricted areas. Something similar happened in the UK during the Second World War, when the penultimate conviction under the Witchcraft Act of 1735 was of Spiritualist medium Helen Duncan, who had informed a Portsmouth mother of her son's death at sea *before* the official Navy confirmation arrived. She was considered a threat to national security.[17]

Cases involving missing people are always emotively charged, and there is also the question of *permission*. If you have not been asked to work on the case, is it ethically sound to be doing so if you don't have permission? In most cases the answer is "no", but there are times when you have to decide that it is 'clearly in the interest of the highest common good to do so', for example if it is a member of your own family perhaps.[18]

Dowsing for lost (inanimate) items is, on the whole, much easier; except when you can't find your keys and are in a hurry to get out of the house – it is next to impossible to remain calm enough to dowse successfully in stressful situations like that. It can be difficult to be objective with personal items, especially those of any value (sentimental or otherwise), and it is often better to get someone else to dowse for you, or at least to help you by corroborating your own dowsing. I've had many successes dowsing for missing keys and such like for others, and my partner once successfully dowsed for a missing file in her office. Using L-rods, she identified the filing cabinet in the archive room, then the drawer, and finally moved her rod along the files in the drawer until she found it. As others in the office had been looking for this file for over a day, they were very impressed that she'd found it in less than half an hour – although she didn't mention that she had used dowsing to find it!

A case study from the 17th Century provides other examples of the ethical pitfalls to be aware of in this sort of dowsing, and it might be good to re-examine this in light of the McCann case. This was possibly the first time in history that dowsing was used in a criminal investigation.

The Dowsing Detective

Figure 13

Jaques Aymar Vernay (Figure 13) was a stonemason from St. Marceline in France. A dowser from childhood, Vernay was accomplished at finding minerals and also lost items and people. At age 18 he located the body of a murdered woman that had been hidden in a wine barrel for four months. His rod twitched when he pointed it at the husband, who promptly confessed to the murder.

In the year 1692, a brutal double murder took

place in Lyon, France. A wine merchant and his wife were slain with a billhook in their cellar, and jewels and valuable had been stolen from a strongbox. Locals were shocked at the violence of the crime, but some remembered Vernay's success in the earlier case and petitioned the King's Procurator to bring him to Lyon to assist in the investigation.

When he dowsed in the cellar, Vernay located several buried items, including the murder weapon. His rod twisted violently in his hands when he walked over the spot where the two bodies had lain, and Vernay himself nearly fainted.

Clutching some items of clothing from the victims, Vernay made his way into the streets around the property, following the twitches of his rod and surrounded by onlookers. He led officials on a trail down the banks of the Rhone, eventually coming upon a gardener's cottage. Inside were the remains of a hasty meal, and Vernay announced that they were looking for three fugitives, who had stopped here and consumed a litre of wine. This was confirmed by the gardener's children, who told them of seeing three men there the day before.

The chase continued along the Rhone valley for over 150 miles, until the rod led Vernay to the small town of Beaucaire and the gates of the local gaol. Here, from a line-up of 13 recently arrested prisoners, he picked out a ruffian arrested only an hour beforehand for theft. Vernay was convinced that he had something to do with the murders, although the suspect denied ever having been to Lyon. The authorities took him back to the scene of the crime in Lyon, whereupon the villain, on being recognised and identified by locals, confessed that he had not been involved in the actual murders, but had been employed afterwards by two southerners from Provence to help carry away their loot.

Vindicated, and accompanied this time by a squad of archers, Vernay resumed the chase and was led eventually to Toulon, where he identified an inn where the two murderers had eaten before boarding a ship for Genoa. The chase was abandoned at this point as the soldiers had no authority outside the French borders and were worried how Vernay's dowsing ability would be perceived by the less-enlightened Italian Inquisition.

The arrested suspect was discovered to be a pirate from Toulon. He was tried before 30 judges and executed in front of a large crowd in grisly fashion by being broken alive on the wheel – possibly the last person in Europe to be executed this way.[19]

Vernay's success led to a surge of interest in dowsing and many people in the area experimented with it, to such an extent that a local priest wrote to his superior to inform him of this "strange practice that seems to have taken over nearly the whole population of Grenoble and the Dauphine". Although many clergymen were enthusiastic about dowsing in general, others thought it should not be used to decide moral issues such as questions of guilt or innocence, as it could lead to injustices.

Vernay was subsequently employed by other police forces on several occasions to track down criminals, but growing public controversy eventually led to him being summoned back to Lyon in September 1692 to undergo testing before a number of witnesses including distinguished physician Pierre Garnier, who published his findings in a 108-page report. In the report, Garnier claimed that 'corpuscles' of 'murderous matter' emitted by the criminals at the time of the offence were left at the scene, where they permeated the skin of the dowser causing a 'fermentation' in the blood that was transmitted through the muscles of the hand to the dowsing rod. A medical colleague present at the scene agreed with this finding, adding the extraordinary claim that, because of their infinitesimal size, the corpuscles could remain in place at the scene no matter how much the air might be disturbed by wind or other causes.

A great public debate followed the publication of Garnier's report, with many letters in the press, books and reports following. This was the Age of Reason; the Royal Society in London had been in existence for just 30 years, founded by luminaries like Isaac Newton and Christopher Wren, imbued with a mission to increase our understanding of the world through scientific experimentation. To the burgeoning Natural Philosophers of the day, dowsing was an ideal area for research and advancement of personal theories. The clergy were not immune either; many prominent Abbés and priests published their own findings; some heavily in favour of dowsing, others denouncing it as a tool of the devil.

Meanwhile, Vernay's popularity with the ordinary people continued to grow, although it seems he was easily led astray into awkward situations of dubious ethical provenance. It was reported in a book published by Father Lebrun that he had been persuaded by a number of folk to dowse along the street of towns and villages to tell whether the ladies of any household had "soiled their honour". This 'dowsing voyeurism', wrote Lebrun, "*soon spread through the town, causing so much slander and calumny and creating such uproar in several families, that the devil had good cause to rejoice.*"

Lebrun also recounted a number of documented errors in Vernay's investigations, and it wasn't long before Vernay was summoned to Paris to undergo further testing. This time a series of six holes were dug in a garden. Four were filled with different samples of metals, a fifth with gravel, and the sixth left empty; the turf was then carefully replaced over the holes. Asked to dowse for the metal-containing pits, Vernay found that the rod reacted only over the gravel pit and the void, but not over the metals. Many contemporary dowsers have fallen foul of similar tests of this nature, usually of the 'dowse a bottle of water hidden under one of six buckets' variety. It is safe to say that this is not a good test of a dowser's ability, especially if performed under stressful conditions, e.g. in a television studio surrounded by sceptical observers.

Vernay was also asked to dowse another murder scene that was a few days old, where an archer sentry had been brutally slain by several sword thrusts, but his rod failed to react over the spot where the victim's body had lain. Perhaps leaning towards the corpuscular theory, a mortified Vernay tried to explain this failure by saying that the rod would not react if the killer had been drunk or angry at the time, or if he had already confessed to his crime.

Following this incident, the disheartened and disillusioned Vernay returned home where he continued to obtain good results with his dowsing, though he was never able to completely re-establish his reputation. From being a national hero, he gradually drifted into obscurity. But far from being a death-knell to further investigation of dowsing, interest continued to grow throughout France as a large number of priors, abbots and curates, even the Bishop of Grenoble, studied the art and practised it themselves. The following year another priest, Claude-François Menestrier, published an essay on dowsing in which he stated that his investigations with other dowsers had convinced him of the efficacy of dowsing, and that there was *'no question to which the rod could not provide an answer'*, although it was *'not very reliable in predicting the future'*. But the prelate was particularly horrified that dowsing should be used to determine guilt or innocence in criminal cases, and his thoughts on this are worth quoting as an ethical and moral postscript to this story:

> *"Let none say that this is a wise disposition of Providence and of God's justice to prevent certain crimes from remaining unpunished. For it is not permitted to justice to utilise directly or indirectly such indications either to absolve or to condemn, inasmuch as they are subject to great error and bad faith on the part of persons who claim they have such a talent. By dowsing with their rods, they can thus render false opinions about individuals they seek to harm. When it is a question of the life, property or honour of persons, one must have proof and testimony."*[20]

Chapter 6
Map Dowsing

The exact dowsing methods employed will depend on the circumstances, but most location dowsing problems can be approached using a combination of **map** and **remote** dowsing. The two terms are often used synonymously, but technically the distinction is that in map dowsing, the dowser is in a completely different location and is working solely from a plan or map; whereas remote dowsing can also apply to situations where the dowser is close to but not at the target area and is using his dowsing tool to indicate the direction of the target. Thus, a water dowser can stand at the edge of a field and ask his rod to show him the direction he needs to walk to find the nearest underground stream. Most people do this by turning round on the spot, holding one rod in front, while asking a question such as, "Which way is North?", or "which direction should I walk to get back to the car?". The rod will stop moving and 'stick' in the direction of the intended target. This method comes in very handy when you're trying to find an ancient site without adequate maps or directions; although in difficult terrain it is better to ask for the most *appropriate* route to take, rather than the most direct route, which could lead straight into a bog or worse. Personal experience in the Forêt de

Dowsing Magic

Paimpont in Brittany with the Earth Energies Special Interest Group bears testimony to this!

In dowsing for missing items or people, often a map dowse is performed first to narrow down the area of search. When the target area is narrowed down sufficiently, the dowser can then go to the area and remote dowse to home in on the target. The first approach is obviously to get a map of the area where you believe the target to be; in the UK an Ordnance Survey map is ideal, and of course online map resources like Google Earth are invaluable these days.

Techniques

There are many methods you can employ for map dowsing and you are sure to develop your own favoured method. Take a look at Figure 14, which shows a site plan of a fairly large area.

Figure 14

Dowsing Magic

Let us suppose that we have been approached by the landowner, who is looking for the best spot to drill a borehole for a domestic water supply. Sit down with the map in front of you[21] and ask your pre-dowse questions with your pendulum to check you have permission on all levels. I add an additional question at this stage, which is: "Can I make a connection with this place (on the map)?"

Assuming all answers are positive, you might begin by narrowing down the search as much as you can.

Having formulated your dowsing question – perhaps in this case a suitable one would be: *"I'm looking for the crossing point of two underground streams that will provide a year-round flow of potable water of at least x litres per hour"* - divide the map in half with a vertical line, then dowse each half in turn with your pendulum, asking "is it in this half?". Continue narrowing it down in this way by dividing the remaining area in half again, and so on. Soon you will have narrowed it down to a small area (Figure 15).

Figure 15

Dowsing Magic

A similar approach can be used if you are dowsing within a building, perhaps for a lost item. Here you can quickly identify which room the target is in by map dowsing a floor plan of the building, then go to that room and dowse with your rods. Of course with multi-storey buildings you will need to use separate floor plans for each floor.

Another approach when looking for point targets is the coordinate method. For this you need a straightedge and your pendulum. Holding the straightedge horizontal at the top of the page, move it slowly down the map whilst dowsing the question. When the pendulum reacts, draw a horizontal line on the map (Figure 16):

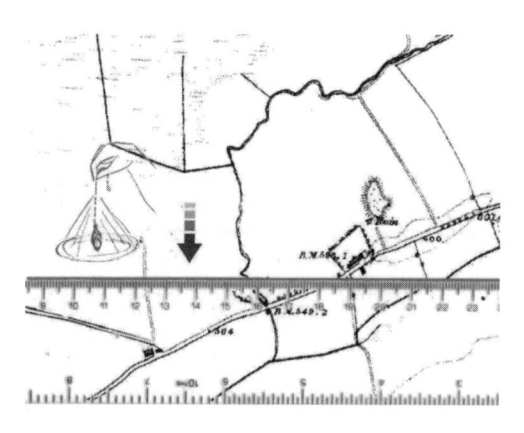

Figure 16

Repeat this exercise with the straightedge vertical and moving from left to right across the page. Very soon indeed you will have a point marked by the two intersecting lines (Figure 17). You can use either of these methods for any point target, be it a lost dog, missing item, or finding the best place for a borehole.

Dowsing Magic

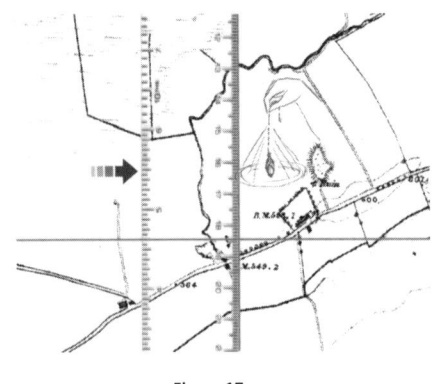

Figure 17

It can be somewhat fiddly manipulating pencil, pendulum and ruler in these situations, especially if you have to keep swapping between pendulum and pencil in your dominant hand all the time whenever you want to write something down. If this is the case, I suggest that you experiment with using your pendulum in your other hand so that you can continue to write with the pencil. I certainly found that my dowsing is just as accurate with the pendulum in my left (non-dominant) hand, although I would still normally use it in my right hand whenever possible as I find it less tiring.

If you're looking for linear features like leys or pipes, an easy method is to dowse along an edge of the plan, moving a pencil along the edge acting as a pointer whilst dowsing with a pendulum in the other hand, asking to be shown

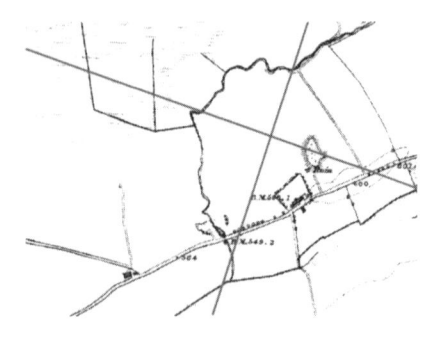

Figure 18

where the line crosses the edge. You can also do this with inverted L-rods, waiting for the rods to cross at the appropriate point, although I find the

Dowsing Magic

pointer-and-pendulum method to be more precise. You can then identify which edge of the plan the line exits, dowse along that edge to find precisely where it exits, and draw a line connecting the two points (**Error! Reference source not found.**).

Alternatively, you can hold your L-rod or pendulum over your first point and ask the tool to indicate the direction of the line; however personally I find this method less accurate.

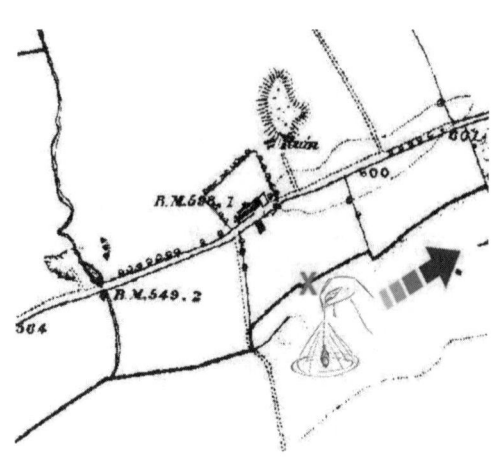

Figure 19

For meandering lines like watercourses it is a little trickier. Find a suitable line at the edge of the area you want to survey, and dowse along it using your pencil as a pointer, just as described above ("Show me where the watercourse crosses this boundary"). Mark that point with your pencil (**Error! Reference source not found.**):

73

Dowsing Magic

Now trace your pencil in a circle or arc around your first point, making another mark when you get your reaction from the pendulum. Continue in this way, marking each point as you go, until you can plot the course of the line across your site (Figure 20).

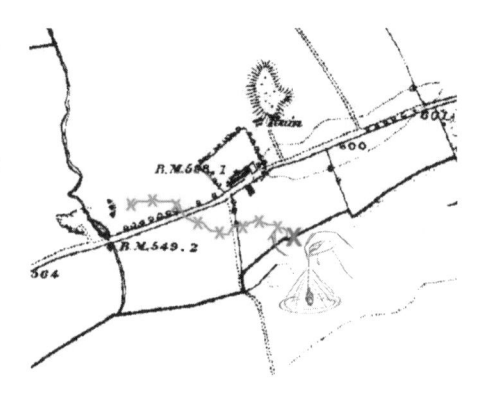

Figure 20

There is another technique that is particularly of interest in archaeology dowsing, and also where you might be looking for multiple targets; this is the

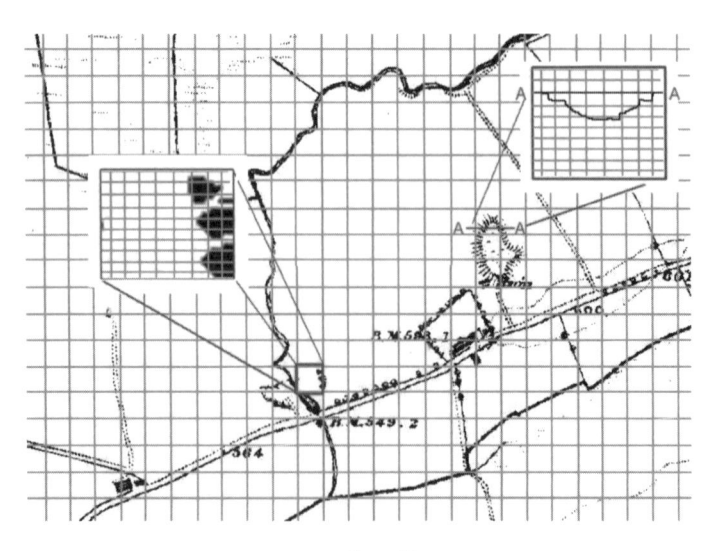

Figure 21

'grid scan' method (Figure 21). By drawing your plan onto fine graph paper, you can pendulum dowse each square of the grid in turn to see if contains anything that you are looking for, e.g. *"human-made structures or artefacts from the Roman period of occupation"*. Every time you get a 'hit', fill in that square with an 'X'. If you find any squares that you want to investigate in more detail, simply enlarge one square onto another bit of graph paper at a 10x larger scale and repeat. You can do this 'scaling up' as many times as you need to get sufficient detail. Sometimes you will find the pattern of your dowsing 'hits' reveal the outline of a structure or artefact. Note that you will probably want to use a finer grid than I've used here – this is just for illustrative purposes. Archaeological dowser Beth Davies has used this method to great success, often identifying buried artefacts like statues or weapons.

It is also possible to dowse vertically through a section of your plan using similar methods. Take the section A-A on Fig. 7 and scale it up until each square on your graph paper will represent something like 10cm in reality. Now you can dowse each square below your drawn ground level to reveal a side view of what is present. You can dowse down as far as you like, but in practice you rarely need to go down more than a metre or two. This method is good for revealing staircases, trenches and such like.

The grid-scan method is an advanced technique. It is incredibly time-consuming and tiring, and it can take several sessions spread over a few days before the map dowsing is completed. Nevertheless, it does have its uses at times and is certainly worth remembering.

These are just some of the methods used in map dowsing. You will most likely use a combination of these depending on the circumstances, or develop some other method entirely with time. Like any new dowsing method, it is important to practise these techniques, and a fun way to do it is to get someone to hide a 'treasure' somewhere in your garden (I find chocolate works well, especially with children). Then, draw up a plan of the garden and practise your map dowsing to see if you can locate the treasure. It is then easy to go outside and (hopefully) confirm your findings using your rods or other favoured dowsing tool.

Chapter 7
Refining technique

More tools

So far we have looked at the main dowsing tools, the pendulum, L-rod, V-rod; but there are many other popular dowsing tools out there. The range and style of pendulums alone is overwhelming; as well as the usual metal, wooden or crystal ones, there are a whole range of 'programmable' pendulums available. These seem to be more popular in Europe, particularly in France, where many of them originate. Some of the more advanced ones utilise the colours of the Mager rosette combined with a moveable ring or pointer so that a particular colour can be selected on the pendulum. The pendulum is then said to be "sensitive" only to this particular frequency. Some are said to "transmit" this frequency as well and can be useful for healing work, either of people or landscape. The most advanced pendulum of this type is probably the 'Universal Pendulum' (Figure 22), which has a moveable gimbal and calibrated string to provide numerous combinations.

Figure 22

Although the manufacturers of these devices may make grander claims for their efficacy, basically what these are doing is providing an easier way to focus the mind on the job in hand. Everything that they do can be replicated using a normal pendulum and some form of witness like the Mager Rosette; and of course if your intent is focused enough even those are not necessary. But they do look very impressive and that does carry a certain amount of weight when you're with a client.

In the rods department, there are several variations on the V-rod and L-rod. I've seen circular rods, U-shaped rods used in pairs, a W-shaped rod with wooden handles and many others. Many dowsers invent their own favourite device, and some of these eventually make it out into the dowsing community at large. Guy Underwood for example, invented the 'geodetic rod' from a screwdriver and a length of stiff motorcycle brake cable, the two connected by a short link so that they can rotate around each other. He did try to market this device as he thought that it was the ideal dowsing tool and that anyone could dowse with it; sadly not very many people agreed with him on that point. Variations of this device were popular with early BSD members who found it useful in situations where counting was involved, such as depthing or identifying sidebands of water veins (of which more in just a moment), but it remains a rather idiosyncratic device and not too many peoples' taste.

Bobbers and Wands

These tools are a sort of 'halfway house' between the pendulum and the rod. They are more stable than a pendulum when used outdoors, and many people prefer them for such use. Basically a long whippy wire with a handle, often with a weight on the end, they combine the yes/no answers or the pendulum with something of the directional ability of the L-rod. Many lengths and variations are possible; I have even seen one made from a tapered carbon-fibre fishing rod body that was about 4 feet long. Both wands and bobbers do basically the same job; although usually a bobber is longer and has a heavier weight on the end so will tend to be used more dynamically than a wand, largely imitating the movements of the pendulum. Figure 23 shows a bobber sold by the BSD that has an additional spring next to the handle. As you

Figure 23

might imagine from the name, wands are used in more subtle areas of dowsing, and are very good at detecting subtle energies. Some people use them as you would a 'magic wand', for the direction and manipulation of intent and energy. There is a rather nice little book on wand dowsing available from the BSD shop if you want to know more about working with these tools.[22]

The Cameron Aurameter

Personally, I'm not a huge fan of either wands or bobbers, as I find both instruments rather bulky and cumbersome to carry around, and limited in their use. However, there is one available 'multi-tool' that combines the movements of pendulum, bobber and L-rod into one device that you can carry in your pocket – this is the Cameron Aurameter (Figure 24). Developed by master American dowser Verne Cameron in the 1960's, this is a bobber *par excellence*. Its sensitivity is adjustable and unmatched by L-rods, yet it somehow manages to be more stable in windy conditions. It can be turned upside down and used as a pendulum by sticking your thumb in the spring; the point can be turned down for map dowsing or turned up and back towards you for any counting task. It is extremely fast to react and so is very good for following meandering earth energy lines or the edges of energy fields. You can also use it as a bobber for counting at the same time as following a line, and it will indicate direction of flow as soon as you hit the line – you don't have to dowse the line twice, unlike L-rods. It was originally

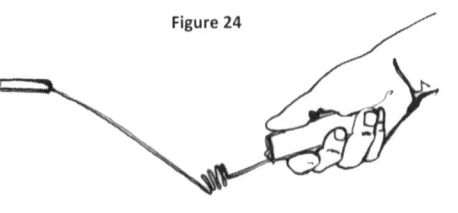

Figure 24

developed for measuring the human aura, as you might guess from the name, and it is very effective at that particular task. It is the most versatile tool I have ever come across, and it does take a bit of getting used to; but perseverance brings rewards. I am still finding new reactions with my aurameter, even after several years of use.

The price tends to put this device out of reach of most people, but it is money well spent as the tool is guaranteed *for life*. That's no ideal boast either - I'm on my second one and the first was replaced within a week with no questions asked and at no cost to me. Where else these days can you get that kind of service?

Depthing and other informational dowsing

Finding a water vein is one thing, but before you can start drilling for it, you will need to know a few more things about it, for example how deep it is, the rate of flow, whether it will meet the client's needs all year round, the purity of the water, and what the geology of the ground is like. The latter can be discovered by a bit of advance research and a geological map of the site, but even with this information it is sometimes necessary to check with dowsing on the site itself if the water is deep and there are many different layers of geology above it.

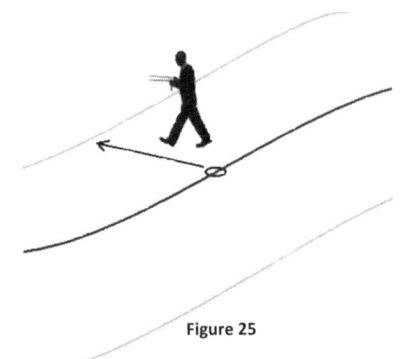

Many water dowsers have their own favourite methods for determining these things, so it's really a question of finding something that works for you. But probably the most well-known method of establishing depth is known as the ***Bishop's Rule,*** named after the Bishop of Grenoble, who discovered the method by observing celebrated French dowser Barthelemey Bléton at work *circa* 1770 (Bléton is perhaps best known for his dowsing of the gardens of Versailles at the request of Marie Antoinette). This relies on the concept that water veins can appear to have *sidebands*. The exact number of sidebands detected varies from dowser to dowser, an indication that what is being dowsed here is not a physical reality but a conceptual one. Some dowsers work with extremely elaborate systems of multiple sidebands and sub-divisions to determine all sorts of subtle nuances of depth and flow rate, but the basic concept of the Bishop's Rule is that you find the centre of the vein then walk outwards from it at right angles until you detect the sideband (Figure

Figure 25

25). The distance you have walked is equal to the depth of the water under the surface.

Another method that was popular in the early days of the BSD was devised by Major R. Creyke and consequently bears his name. In **Creyke's Method,** a stake is driven into the centre of the water vein. This has the effect of making the surface line apparently disappear downstream of that point, and instead a circle manifests, whose radius is equal to the depth of the stream (Figure 26).

A refinement of this technique provides for two concentric circles as in the diagram, and the distance between the two equates to the distance between the top and bottom of the water layer.

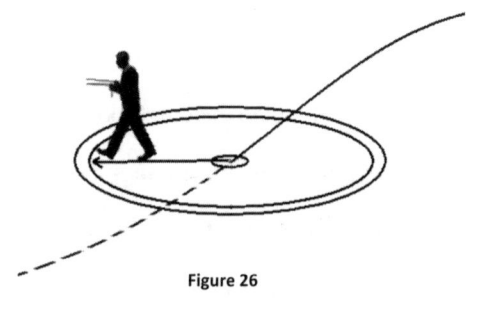

Figure 26

Both *Bishop's Rule* and *Creyke's Method* have limited accuracy over certain types of geology. Clay in particular seems to confuse things, and water dowsers learn through experience to add or subtract a percentage from their dowsed depth when working in these areas. Most will also confirm their findings by using another method. One that was popularised by Major Creyke and other early BSD members was to use a customised rod similar to Underwood's geodetic rod where the rotations of the rod over the water would equate to its depth in feet. You can also use a pendulum or even a Y-rod for this task of course, indeed you can use a pendulum for the whole depthing exercise simply by asking; "Is this water more than 100 feet deep?" (no) "Is it more than 50 feet?" (yes), and so on until you arrive at a number. Imagine yourself mentally scanning downwards through the ground rather like the sectional map dowsing exercise we learned previously.

If the geology is varied, you can refine this by counting aloud with the pendulum swinging in search position, asking it to change direction or otherwise react at each change of geology, for instance you might dowse changes at 10 feet, 14 feet, 23 feet, 38 feet etc. until you reach the depth of the water. If you have a good working knowledge of the local geology you can also dowse to see what each layer consists of, e.g. topsoil, clay, gravel, water-bearing sand etc. This information will be valuable to the borehole driller and will enhance your reputation as a reliable water dowser.

Counting techniques like these can be applied to a variety of dowsing situations where you need to quantify something. In many cases the result will be a completely subjective number that is only meaningful to you. For example, I once had the opportunity to dowse the 'energy' of a performance area in Tokyo before and after it was blessed by a Shinto priest. I asked if there was a 'unit scale of spiritual energy' that I could dowse, and got a 'yes' to this question, so I then dowsed using the counting technique. The 'before' energy dowsed at around 84 'units', and after the blessing ceremony it was something like 323, rising to a peak of 430 about an hour later. This information was meaningful to me, as it showed a change in energy level over time; yet I would not expect another dowser to come up with the same figures as I did as the 'energy units' might be different in their particular picture. I would however expect them to show a similar trend in the rise and fall of the energy.

Over the years, several devices have been produced that attempt to provide an objective measuring scale for dowsing, such as the Bovis Biometer. This gadget was devised by a French physicist Antoine Bovis as a means of measuring what he called 'life force' energy. In its simplest form, the Biometer is a small box with a sample container on the top and a graduated sliding rule in the base. A sample is placed in the container, and the pendulum is swung in the search position over the rule, which is then gradually slid in or out of the box until a reaction is achieved. The reading is noted and becomes what is called the 'rate' of the sample. A simpler technique uses circular numbered charts that the pendulum is held over.

At its most basic level, the gradations on the Bovis rule range from 1-100, but more complex scales ostensibly based on Ångström unitsve been used, with a 'neutral baseline' of 6500 Ångströms (above that reading is beneficial to life, below it is detrimental); although because there is no exact correlation and to avoid any confusion, they are referred to simply as Bovis Units.

Varying ranges and variations on the Bovis scale have been adopted by different dowsers over the years, and this highlights the main problem with such systems; namely that readings tend to be highly subjective to the individual dowser, and not the universal scale of measurement that their creator hopes.

Gadgets like the Biometer tend to be more popular with European dowsers, although many health dowsers use similar measuring devices. These devices also led to the development of the controversial field of radionics, where complex gadgets are employed not only to gauge the 'rate' of a sample, but also to 'broadcast' the dowser's intent. There are circumstances where it might be useful to incorporate such props into your dowsing, but don't let yourself become dependent on them; they may not be there for you when you need them. A far more pragmatic attitude is to always be able to improvise any dowsing task with only the materials at hand.

Witnesses

In the 'props' category, we need to also include 'witnesses'. Strictly speaking, a witness is a sample of the target being dowsed for that is held in the hand or incorporated into the dowsing tool in some way, such as a small bottle of water taped onto the end of your Y-rod, or a small piece of gold held in a chamber in a pendulum. The witness acts as a focus for the mind, and the idea is that you 'attune' to and only get a dowsing reaction from whatever that substance is. Many dowsers don't feel 'right' unless they use a witness, even though they might ridicule it as superstition. But always remember that it is just a mental tool – you don't need a witness to be able to dowse successfully. It should be enough to hold a focused mental picture of what you are looking for.

Taken to extremes, you can find some archaeological dowsers using a whole barrage of witnesses to identify substances like tallow, ashes, human and animal bones, iron, oak, linen and so on. Sometimes a very detailed picture of a site can be dowsed using these methods, even before any

Dowsing Magic

excavation has been undertaken. Yet without corroboration from the actual excavation, the most advanced dowsing picture has no value in archaeological work. It's no good building up an elaborate picture of underground structures using your collection of witnesses if the actual excavation does not support your findings. Still, archaeology is a very rewarding discipline to work in as a dowser if you can find a sympathetic team to work with, as it is one of the few areas of dowsing where you can expect to get early confirmation (or not!) of your dowsing findings.

Here, once again the caveat is not to become dependent on your selection of witnesses. There will come a time when you leave your sample library at home, and you do not want to get to the site and be unable to dowse because you don't have your props with you. If you ever find yourself in such a situation and feel that you really need a witness, try simply visualising the sample witness as you would normally use it; or even write the name of the witness on a piece of paper and hold that – pretty soon you will realise that they are just mental aids and that you can do without.

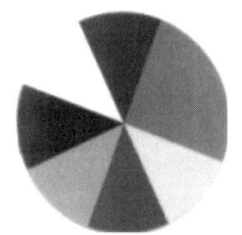

Figure 27

Another sort of witness that we have mentioned earlier is the Mager Rosette, an 8-sectored disc of white, black, grey, red, yellow, green, blue and purple segments. (Figure 27). This might be thought of as a 'passive' witness, as it is most commonly used to ascertain the character of a dowsed target (such as a water line or minerals) *after* it's been dowsed. To use it, hold one of the coloured sectors between two fingers of your free hand, and use your pendulum whilst you mentally ask if the target 'resonates' to the colour you are holding. Move the disk round between your fingers until the pendulum reacts to a colour. That tells you the 'flavour' of the target. For example, common attributes for water are white=potable; grey=polluted; purple=energised or 'holy' water; yellow=sulphurous; red=chalybeate (containing iron), and so on. Black means energetically polluted, in other words carrying geopathic stress, and when you hear a dowser talking about a 'black stream', this is generally what they are referring to.

Colour systems can be used to categorise many things beside water, and can also be useful in healing to dowse the colour of the aura. As colour symbolism tends to be a personal and subjective affair, most people will develop their own system of correspondences over time, and may incorporate more colours than the Mager disk uses. For instance, 'metallic' colours like copper, gold and silver can be useful for earth energy work and are equally useful for mineral dowsing.

You can of course also simply mentally visualise each colour in turn as you dowse them with the pendulum instead of using the Mager disk, and you will

Dowsing Magic

probably develop this ability naturally over time as your dowsing skills mature. If you have the psychic ability to see auras then you will have no trouble with this exercise. Otherwise, have faith in your ability and keep practising – it will come to you eventually.

Chapter 8
Introducing Earth Energies

Up to now we have barely mentioned dowsing for intangible targets such as earth energies. Perhaps it's appropriate that we have left this for the later chapters, as it is only in recent years that this form of dowsing has matured.

Our understanding of the whole 'earth mysteries' scene has similarly developed since the early days of 'New Age' thinking, with its talk of ley lines, UFO flight paths, spirit roads, 'earthlights' and so on. Researchers like John Michell, Paul Devereux, Tom Graves and many others too numerous to mention have all helped shape our conceptual model to the point where in more recent years it has become perfectly acceptable to talk about such topics in respectable circles.

As our understanding of earth mysteries has grown, so has the area of earth energy dowsing, which became the most rapidly expanding area of research in the British Society of Dowsers, leading to the formation of the Earth Energies Group in 1994 – the oldest of the Society's Special Interest Groups. Largely due to the research endeavours of the EEG, concepts like Watkins' leys and the 'dragon lines' of Feng-Shui. have been integrated with more modern scientific concepts like string theory, torsion field theory and spin theory, which looks at the effect of solar system orbital mechanics on the Earth's energy field, to give us a pretty comprehensive picture of the sort of energies that dowsers find. There are energy leys, global geomagnetic grids, planetary grids, toroids, vortices, sinuous dragon lines and a bewildering number of other energetic manifestations that can be dowsed. This gives us an insight into what is perhaps the main problem with earth energy dowsing – the fact that our consciousness, influenced by our individual worldviews and prior learned experiences, plays such a large part in interpreting what we find.

So it is no surprise to find that people can construe things differently when it comes to earth energies[23]. It is vital therefore, that dowsers working in this field compare findings and work together whenever possible to build up a common vocabulary, and this was (and still is) one of the main purposes of the EEG. Over the early years of the EEG a large body of knowledge was accumulated and disseminated in the EEG newsletter, *'Earth Energy Matters'*. Founding Chair Billy Gawn later collated the *EEG Encyclopaedia of Dowsing Terms,* now in its third edition, which aimed to provide a common frame of reference for earth energy dowsing. Both the newsletter archive and the *Encyclopaedia* are invaluable resources to novice and veteran dowsers alike.[24]

Another problem is that dowsing earth energies can be a very interactive process - just as they can have an effect on your consciousness, so your consciousness can have an effect on the energies, changing what you (and others) find. The brain is like a computer – it's just that some of us are Macs and some are running Windows. We each have to learn how to program our individual biocomputers, and advancing in our dowsing provides us with the tools to take control of our own operating systems. Dowsing has tremendous potential as a tool for personal and spiritual growth, and this is an area where I expect to see a lot more development in the future.

We'll look at these aspects in more depth later, but for now just bear in mind that although I will try to give a general picture of earth energy dowsing in this section, what follows is largely my own 'map of the territory'. Nonetheless, it helps to have a baseline from which to start and hopefully you will find it useful in that regard.

Definitions and terms

So what exactly do we mean by *earth energies*? It's such a wide-ranging term and has been subject to much misinterpretation over the years. But basically, any energy that can be dowsed for **on** planet Earth can be classed under the heading, even although those energies may not originate **from** the Earth. Thus, it includes etheric fields from animate and inanimate objects, human auras, thought forms, planetary influences, geomagnetic fields and so forth. We really should just be talking about 'energy dowsing', but in many ways that term sounds even woollier; so we're stuck with 'earth energies' for now.

The term 'ley line' is another misnomer that has now become enshrined in common usage. 'Ley' is an archaic English term meaning a cleared straight strip of land, and was first used to describe an alignment of ancient sites, old churches, moats, crossroads and so on by Alfred Watkins in the 1920's because he found so many place names containing the term along the routes of his alignments. Since ley already means 'straight', semantically it is a tautology to add 'line' to this. Watkins himself never used both words together, and even tried to stop using the term 'ley' in later years, coming to prefer 'archaic track' instead. Actually, Watkins wasn't the first to postulate long-distance alignments of ancient sites; according to Wikipedia, in 1870 a gentleman named William Henry Black gave a talk to the British Archaeological Association in Hereford, in which he stated that *"Monuments exist marking grand geometrical lines which cover the whole of Western Europe"*[25]. Later, pioneering archaeo-astronomer Sir Norman Lockyer discovered the Stonehenge-Old Sarum-Salisbury Cathedral ley shown here (Figure 28).

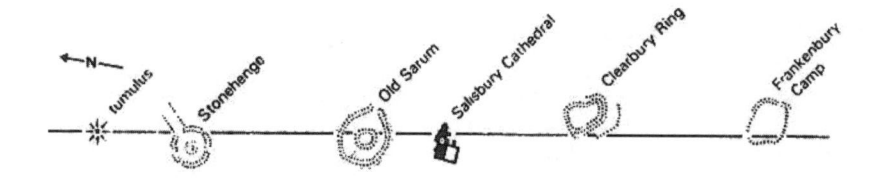

Figure 28

For dowsers the term 'ley' is doubly confusing, as not all leys are the same - some of them seem to be purely visual alignments of sites. Whilst these *can* be dowsed, the lines are usually very narrow and contain little or no energy, indicating that they may be purely conceptual in nature, or perhaps that the energy they once contained has been disrupted or diverted elsewhere.

For dowsing purposes we are mainly interested in another sort of line, which *sometimes* corresponds with a visual ley, but can also be found independently. These dowse as wide straight bands of energy flowing on or just above the ground surface. The grandfather of earth mysteries research, the late John Michell, suggested that we use the term 'energy ley' to distinguish these straight energy lines from 'normal' leys, and I try hard to adhere to that distinction.

Dowsing the leys

In addition to leys (both energy and visual), there is a bewildering variety of other energies that we can dowse for - you can find sinuous lines of earth energy, spirals, vortices, auras, toroids, several varieties of grids, and even geometric shapes and pictograms. We'll look at some of these later on, but for now let's keep it simple and concentrate on energy leys, as these are relatively easy for the beginner to find. If you think of water as the 'yin' of the earth energy world, then energy leys are the 'yang'. They're like espresso coffee compared to water's green tea.

Energy leys seem to originate and terminate at *blind springs* (which we looked at in Chapter 4) and so are frequently found at sacred sites such as stone circles and in older (generally pre-Reformation) churches, which were often built on much older sacred sites[26]. In a church, you will typically find a blind spring under the altar, and very often an energy ley running down the axis of the building. A site such as this is the best place to start when dowsing for energy leys although, since not all church users are tolerant of dowsing, a stone ring is probably a safer choice to begin your explorations.

In a typical stone ring, you will be able to dowse a blind spring inside it, although not necessarily in the exact centre. Using our direction-finding technique, stand with one rod and turn around on the spot, asking 'show me the direction of the blind spring here'. The rod will stick and point in that direction; you can then walk towards it until your rods cross. If it's a strong dome, remember that your rods may even start to whizz round like a helicopter. When you've found the blind spring, dowse the extent of the 'dome'. This frequently coincides with the diameter of the stone ring. Next, walk around the perimeter and dowse the number of water veins exiting the dome. Find the veins first, then find the direction of flow with your rods. You should also find at least one stream *entering* the dome, which will be the feeder stream for the system.

Now that you know where the water is, try to find any energy leys. Walk around outside the perimeter of the water dome once more, telling your rods that you want them to indicate any energy leys. Visualise these as straight lines of glowing white energy flowing across the ground. You can dowse for them in the same way as a water line - find the centre of the line first, then step back a few paces and ask for the 'nearside edge', then the far edge. You should find that the width of the energy ley is approximately 2.5 - 3m, although they can sometimes be much wider. You may find them narrowing down to a smaller 'node' at the blind spring if there is a crossing of leys.

Although energetically an energy ley feels 'yang' to the 'yin' of a water line, the ley energy is much subtler so don't expect to get as strong a reaction as you would from a water line. Initially it may take you a couple of passes to be sure that you have found one. I find that for leys, L-rods will swing *outwards* instead of crossing as they do with water, and you can of course mentally program your rods to do the same for you with a bit of practice. This technique is very useful as it allows you to dowse both water veins and energy leys in a single pass of a site. Pay attention to bodily signals too – I find that my feet will become warm or even itchy when I'm standing on an energy ley, whereas with water I feel it more in the pit of my stomach. Bodily signals like these are a useful confirmation of your dowsing, and cultivating these is the first step to becoming a deviceless dowser. You can also try crouching down and placing your hand into the energy ley about a foot above the ground surface; the air should feel noticeably warmer.

Having completed your circuit of the site, you should have found at least one ley, and usually not more than 3 or 4 unless you are at a major power centre. If you find an odd number, it means that one of the leys is originating or terminating at this power centre. You can determine which by dowsing for the direction of flow with your rods, just as you did with the

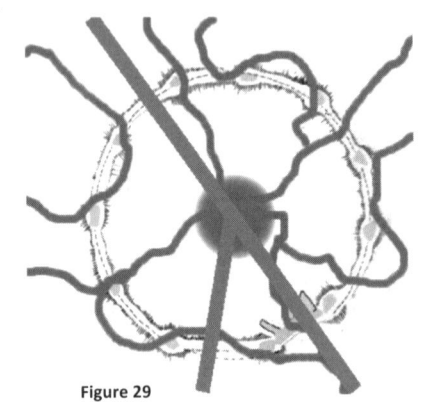

Figure 29

water.

Usually an even number of leys will indicate a crossing point (although of course they have to be directly opposite each other across the dome for this to be true). The picture (Figure 29) depicts a hypothetical example of water flows and two energy leys at a stone circle, one passing through and one either originating or terminating there. Note that it's easy for novice dowsers to confuse water reactions and energy leys. If you find more than five energy leys running through a site, you've probably become muddled somewhere. This is why it's good to have a more experienced dowser on hand to confirm (or refute) your findings.

Character of energy leys

Just as we did with water, you can use colours to classify the energy of a ley using the Mager Rosette. The colour attributes may be different for you with leys, so you will have to learn from experience what each colour means. Some commonly used attributes are purple for 'holy lines' in churches, white for 'normal', black for unhealthy lines and so forth. As I mentioned earlier, it can also be useful to apply a wider range of colours as the Mager disk is a bit limiting in this regard – you may want to try using metallic colours such as gold, silver or copper. However, it may take several years of experience before you are confident and consistent with your personal colour system, so I would suggest that initially you use colour solely as a means of differentiating one line from another, without assigning any meaning to it.

An additional classification that you can apply is to dowse whether the energy is positive, negative or neutral; thus you might hear some dowsers talk of 'negative green' energy (generally thought to have a preservative effect). You have to be careful when using terminology like this, as many people perceive 'negative' in a pejorative way. Think of it like electricity; that can be positive, negative or neutral too, yet we don't think of negative as 'bad' in an electrical context. More useful terms to apply to energy leys might be 'healthy/traumatised', 'stressed/balanced', or 'beneficial/detrimental'.

Energy leys are built up from smaller, thinner lines of energy, and you can dowse these by making a slow pass over the ley, resetting your rods to the search position after each band. There always seems to be an odd number of bands that flow in opposite directions and are of opposing polarities. The majority of the bands will determine the overall polarity and flow direction of the ley, as shown (Error! Reference source not found.). Usually the colours of the individual bands will be reflected symmetrically across the whole width of the line; for instance if your outermost band is green, the one on the other side will be green also.

In certain circumstances, these bands can 'delaminate' for a short distance, before coming together again. I have noticed this in some church leys, where one or more of the outer bands left the main flow temporarily to detour through the font or some other feature.

−ve +ve

overall direction of flow →

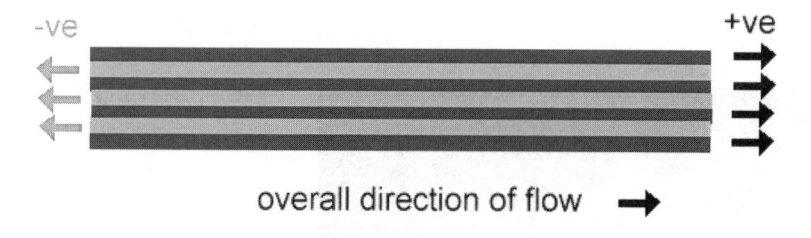

Figure 30

You can also find what appear to be individual bands in isolation, unconnected with energy leys or sacred sites. These narrow 'energy lines' can be found meandering through the landscape, perhaps connecting with natural features, forming spirals and other more intricate patterns as they interact with themselves and their surroundings. I consider these to be earth energy in its most 'wild' or 'untamed' form, unlike the more formalised energy leys where several energy lines seem to have been wrought or channelled together to work as a whole – although whether this is a natural phenomenon or indicative of human intervention is still a matter for debate.

The bigger picture

Energy leys are usually dowsed on a local level and for short distances – but where do they come from? What happens in the larger landscape picture? Is there an overall global pattern to them?

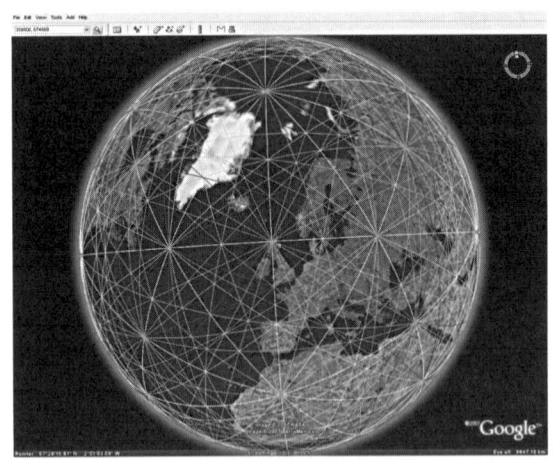

Figure 31 © Google/ NASA/ TerraMetrics

The concept of a global energy grid has been around since the early 1970's, when a group of Russian scientists proposed a global energetic grid based on the geometric form known as an icoso-dodecahedron. The geometry of the grid was supposed to determine the etheric pattern of the planet, determining the placement of significant geographical features; and there did appear to be some correspondence with the edges of tectonic plates.

In the 1980's, US scientists Bethe Hagens, a Professor of Anthropology at Governor's State University and William Becker, a Professor of Industrial

Design at the University of Illinois, Chicago, developed the earlier theory to include the other Platonic solids – tetrahedron, cube, octahedron, and icosahedron; plus a couple of other regular geometric figures with even more exotic names – the rhombic dodecahedron and the triacontrahedron. They called the finished grid the 'Unified Vector Geometry Earthstar 120 grid' – usually abbreviated to 'UVG grid'. Prof. Hagens has recently produced a Google Earth placemark of the grid, which you can see in the illustration (Figure 31)[27].

The work of two other scientists also needs to be taken into consideration here. Back in 1979, scientist James Lovelock advanced the theory that the Earth was a living being in its own right, known to the ancient Greeks as *Gaia* – and that humans and all other life forms on the planet are a part of the same living system. More recently, biologist Dr. Rupert Sheldrake proposed the concept of the *morphogenetic field,* an energetic blueprint that determines the growth and form of living organisms – perhaps better known as the *aura* to most of us.

These theories have profound implications for our perception of the landscape for if the Earth is indeed a living being, then it seems reasonable that she too will have a subtle energy body that not only determines the distribution of major geographic features, but also has chakras (energy nodes), meridians and energy flows that mirror those found in the human body and, importantly from our point of view, can be dowsed.

Enter the dragon

If you're familiar with the work of John Michell or Hamish Miller you will know about the Michael line, but you may not know that this corresponds to one of the lines on the UVG grid. The Michael line is a 'geomantic corridor' of sacred sites across southern England, running just north of St. Michael's Mount and then through places including the Cheesewring, Brent Tor, Burrow Mump, Glastonbury Tor, Avebury, Royston and Bury St. Edmunds, many of which have an association with St. Michael. The line is aligned on the direction of the Beltane sunrise, and the fact that there are so many significant natural landscape features along it lends credence to our hypothesis that the world grid determines the geographic placement of such features. Hamish Miller and Paul Broadhurst spent some time dowsing along this alignment, documenting their findings in *The Sun & the Serpent*. The pair later dowsed another long distance alignment, the Apollo line, across Europe, and wrote up their findings in *The Dance of the Dragon*. This line also connects several significant landscape features such as Skellig Michael off the south-west coast of Ireland, St. Michael's Mount, Mont St. Michel in France, Monte Gargano in Italy, Corfu, and the Greek oracle sites of Delphi and Delos. Many of these sites are also dedicated to St. Michael.

Hamish found that these lines were not simply straight flows of energy; they had serpentine *yin* and *yang* components to them, which wound back and forth across the nominal straight alignment, rather in the manner of snakes coiled about a caduceus. Often they crossed at a sacred site such as a church or well, where they formed a node point. The lines would seem to narrow down almost to a point at the crossing, as though they were disappearing into (or emerging from) the earth. He found that the masculine, or *yang* lines tended to go through places of patriarchal power like churches or high points in the landscape, whereas the feminine, or *yin* lines had a tendency to favour more Goddess-oriented sites - churches dedicated to female saints, holy wells and low points in the landscape. He named the two serpentine currents *Michael* and *Mary* in the UK example, and *Apollo* and *Athena* in the European example. These 'dragon lines' have long been familiar to Chinese geomancers, where they are known as *'lung-mei'* or 'dragon's breath'.

There is another long-distance alignment in the UK that was postulated by Guy Ragland Philips in his book *Brigantia: a Mysteriography,* and subsequently researched by Gary Biltcliffe for his book *The Spine of Albion*[28].

Running from Inverhope in Scotland down through St. Catherine's Hill in Winchester, this is known as the Belinus line and its two component serpent lines are named *Belinus* and *Elen*.

When we look at these three leys in Google Earth in conjunction with the UVG grid (Figure 32), we immediately see that there is some correspondence between them, although each ley is displaced from the corresponding UVG line by around 10 degrees clockwise. Perhaps the UVG grid isn't entirely accurate, or maybe it's more of a Platonic ideal of a grid and things have moved somewhat over the years through tectonic plate movement or other causes - but leaving this discrepancy aside, the fact that there is *any* correspondence with the UVG is possibly the best confirmation I've seen that there is some validity to the world grid theory. Our planet really does seem to have a beautiful crystalline energy body that dictates the placement of sites along significant energy flows.

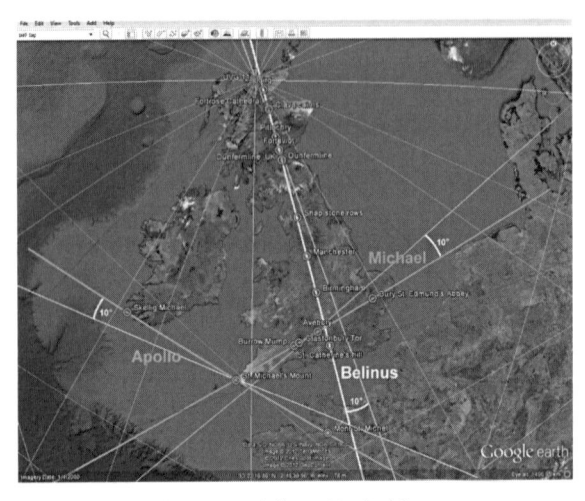

Figure 32 © 2012 Google/ TerraMetrics/ Geocontent

Summary

Each of these major grid lines has very strong serpentine flows associated with them and node points are found where the lines cross, which usually occurs at a significant sacred site. The dragon currents are of opposing polarities, yet they are complementary and interdependent on each other. A good way to picture these is to try and picture things three-dimensionally. Think of a vibrating string or perhaps a skipping rope with nodes and standing waves along it; the polarised lines may just be different 'sides' of the same standing wave.

With this model, an energy ley at a local level can be seen as just a lower vibration of this same global grid harmonic. You *can* dowse serpentine currents associated with them, but the lower vibration means that the serpent lines are contained *within* the energy ley and cannot be detected outside of it.[29]

To continue our earlier electrical analogy, the ley system connecting stone rings and sacred sites is rather like our modern electrical power grid, with local substations stepping down and distributing the power from the high-voltage long-distance transmission grids. The man-made ley network of ancient stone rings and sacred sites may be doing just that, harnessing and distributing the energy of the global grid to areas where it is needed.

Chapter 9
Life, the Universe, and Everything

In the last chapter we looked at the global UVG grid developed by William Becker and Bethe Hagens, and its relationship both to long-distance alignments such as the Michael and Apollo lines. We also looked at its relation to the localised energy leys and other lines that dowsers find. But what is the nature of these lines? Are they related to the Earth's magnetic field, or is there something else going on?

We know that energy leys are affected by the Sun and, to a lesser extent, the Moon. Dowsing during total solar eclipses has shown that the leys and grid lines will either contract almost to nothing or expand exponentially, before springing back like an elastic band afterwards. At the equinoxes, lines typically expand and then contract again afterwards[30], whilst at the solstices, the entire earth grid system seems to collapse completely for about an hour before 'rebooting' gradually over the next day or so[31]. Something similar can be dowsed at sunset, as during the night many earth energy lines will become barely detectable or even disappear completely. This is our clue that there may be an electromagnetic component to the lines, as corresponding changes take place in the Earth's magnetic field at these times.

Dowsing Magic

It's all done with magnets, you know

The Earth is like a giant generator and produces a strong magnetic field around itself, known as the magnetosphere. It is this field that deflects much of the damaging cosmic radiation that would otherwise penetrate the atmosphere and cause all sorts of problems for life on Earth. Indeed, life on Earth would almost certainly be impossible without this protective magnetic umbrella.

Although technically the magnetic field is pretty uniform, it is commonly depicted by a series of loops running between the two magnetic poles, looking rather like a toroid or doughnut. Remember the physics experiment at school where a bar magnet is placed underneath a bit of card and iron filings are scattered on top of the card? The iron filings arrange themselves in definite lines along the strongest areas of magnetic flux, and the same thing happens with the Earth, as shown in this supercomputer simulation of the earth's magnetic field (Figure 33). During some displays of

Figure 33
(NASA/JPL)

aurora borealis (or australis if you're in the southern hemisphere*)* you can actually see the glowing particles form along the lines of higher field strength.

The part of the Earth facing the Sun (i.e. during daytime) also takes a battering from the solar wind, a plasma of highly energetic charged particles ejected from the Sun's corona. When these strike the Earth's magnetosphere, a bow wave is formed and the magnetosphere is compressed, tightening up the field lines. On the side facing away from the sun (i.e. at night), the reverse happens; the field strength becomes weaker as the geomagnetic field is stretched out like a teardrop or comet's tail (Figure 34). As the Earth rotates and dawn approaches, the field is once again compressed and the flux lines tighten up, producing a surge in field strength.

Figure 34 (NASA/JPL)

The field is at its weakest between roughly 2 and 4 in the morning, which is why many people will wake up during the wee small hours, as the body senses the fluctuating geomagnetic field. This is also why one should ideally align the bed so that one's head is to the North for the best night's sleep. In this position, the direction of flow is the same as that of the body's own natural field; if you sleep E-W, then the flow of the geomagnetic field is cutting across your own at right angles.

It is known that the geomagnetic field produces 'telluric currents' in the surface layers of the Earth that are detectable by scientific instruments. Thunderstorms also induce these electrical disturbances. These currents have diurnal characteristics wherein the general direction of flow is towards the sun; although they will move towards the equator during daytime and towards the poles at night. Analysis of disturbances in these currents is used in geological prospecting methods and modern archaeological surveying. They were also the basis of Nikola Tesla's long-distance power transmission theories, and of course in our modern electrical distribution system where the neutral wire is connected to earth at the substation, any leftover current is simply sent through the earth back to the power station. Little is known about how 'dumping' this excess electricity into the ground affects the natural currents, and in turn living organisms; but Gary Skillen, a Canadian dowser who specialises in agricultural work, has reported cases of cows getting severe electric shocks from the ground during stormy weather conditions where there had been a localised build-up of charge[32], and the detrimental effects of stray voltage on milk production of dairy herds have also been well documented by researchers[33].

Changes in geology where an area of ground containing metallic ores or holding a lot of water adjoins an area that is less metallic or dry produces what is called a 'conductivity discontinuity' boundary, and this will have an effect on the daily variations in the telluric currents, either reinforcing them or dampening them. Many ancient 'sacred' sites are known to be located over these conductivity discontinuities, implying that our ancestors were aware of and were deliberately manipulating these telluric currents.

The banks and ditches of henges will actually channel the telluric currents around the outside of the space, only allowing them to pass through the entrances to the henge. Dolmens and stone chambers similarly contain and amplify the energy and produce very interesting toroidal vortices that can be dowsed. Many stone rings even contain 'special' magnetic stones, easily identified with a cheap compass, and these of course will have an effect on the energies flowing through them that can be identified and traced with your dowsing rods.

There is a fascinating book by John Burke and Kaj Halberg called *Seed of Knowledge, Stone of Plenty*[34] that explores these areas in depth, and the authors have done several experiments showing that crop growth is dramatically enhanced if seeds have been exposed to these energies before germination, using stone chambers in New England, temples in South America, stone rings in the UK, even Egyptian pyramids. This book is one of the most important pieces of research of recent years and, although I'm not entirely comfortable with the idea that the Ancients constructed these spaces primarily to enhance crop fertility, it does demonstrate pretty conclusively that the spaces were built where they were because they wanted to exploit the unusual magnetic characteristics of the area in some way.

The European model

So far, it seems that our conceptual dowsing model of energy leys and the world grid fits in pretty well with this geomagnetic picture – or does it? Perhaps it's not the whole story. There is an alternative way of picturing and dowsing these telluric currents that was developed in Europe and was only introduced to the UK about 30 years ago; and – surprise, surprise – it's another series of grids.

Two well-known global grids discovered by German researchers in the years following WW2 are the Hartmann and Curry grids (Figure 35). It is thought that they are connected with the telluric currents and therefore have an electromagnetic component to them; although they seem very difficult to detect using conventional scientific instrumentation. They do seem to vary in strength around the solstices and equinoxes, not to mention eclipses, and the

Dowsing Magic

BSD 'resident' dowsing physicist Jim Lyons thinks that they may result from wave interference patterns generated by the complex interaction of the Sun and planets.[35]

The spacing of the grid lines varies with latitude, but in the UK the Hartmann squares are around 1.6 x 1.7m apart and oriented north-south, with the Curry being slightly more rectangular at 2.4 x 2.7m and offset at 45 degrees. Individual grid lines are around 15-30cm wide. Every so often on the grids, you get a double-strength line that can be either negative or positive in polarity; with the Hartmann, these occur every 35m or so, and with the Curry every 50m or so.

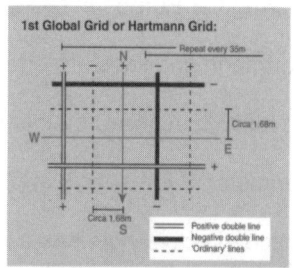

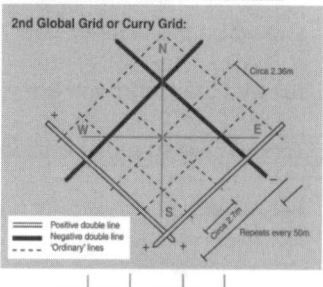

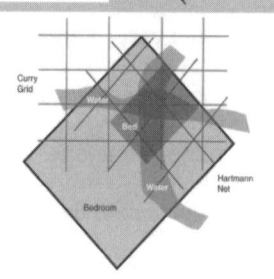

Both Hartmann and Curry grids vary in intensity during the day and are often stronger at night. They are also distorted by nearby power lines or other strong electromagnetic fields, and also by the presence of geopathic stress. Of the two, the Curry grid seems to be more detrimental in this regard, and a double-negative Curry crossing in combination with underground water veins, as shown in the illustration, is thought to be very harmful to human health.

Another European researcher, Blanche Merz[36], has noted that the Hartmann grid lines are often completely excluded from sacred sites like Egyptian temples and older churches, with the lines bunching up around the perimeter of the space; implying that some aspect of the construction of the space effectively creates a neutral grid-free pocket. It is not clear why this should be the case or how it was done; we can only assume that the builders felt that the Hartmann energies are not conducive to the functioning of the space. Like so much ancient geomantic knowledge, this is something that has been lost.

I had a very interesting confirmation of the absence of the Hartmann grid when dowsing in St. Serf's church at Dunning, Perthshire. The building was constructed in at least three phases, with the original chapel built by St. Serf being replaced in the early 13th Century by a tower and church building. In the 17th century, the chancel at the east end was replaced by a laird's gallery and the altar moved to the west end of the building. Finally, in the early 1800's the whole building was enlarged, a pulpit placed centrally on the south wall, and a new aisle extended out from the north wall, giving the building a T-shaped ground plan. During this work a magnificently-carved Pictish cross-stone, the Dupplin Cross (Figure 36), that once stood on the north side of the River Earn

Dowsing Magic

overlooking Forteviot was relocated and displayed in the base of the tower. St. Serf is reputed to have slain 'a great dragon' here when he founded his church, and the cross has a rather splendid dragon carving on the side, indicative of the dragon energy of the male *Belinus* current that still flows through it today in its new home.

Because the last phase of major alteration was done long after the Reformation, it would appear that whatever geomantic knowledge that went into the original construction had been forgotten. Whilst power centres can be dowsed at both the east and west ends of the church, with a connecting energy ley, there doesn't seem to be much happening under the newer pulpit on the south wall. The Hartmann grid, which is easily found in the churchyard, is absent from the older part of the building, with the lines bunching up tightly at the entrance. Walking down the north aisle into the newer part of the building, another thick bunch of lines appears where the old boundary wall stood, and the Hartmann grid reappears under the newer extension.

Figure 36

All in all, a fascinating building to dowse and I do recommend stopping off if you are passing the area; it's not far off the main A9 road a few miles south of Perth and is clearly signposted.

To summarise, both the Hartmann and Curry grids are relatively easy to dowse, and provide an alternative way of looking at the earth energies. They are like the warp and weft in the weave of the Earth's energy field. I don't tend to focus on them as the primary means of assessing a space, but they can be useful in filling out the overall picture and providing extra detail in a consultation.

Since the discovery of the Hartmann and Curry grids, a bewildering array of other global grids has been identified, many with exotic names like Angel, Dragon, Merlin, Yod, Reshel and so on. Billy Gawn dowses grids from the nearer planets and reports that *all objects* have a grid-like energy field[37]. Researcher Gunther Schneck has identified some two dozen distinct earth grids, and shown that they have a close connection with the human energy field and consciousness.[38] Each grid has its own spacing, frequency and other characteristics to be explored.

A wee bit about geopathic stress

Like water lines, energy leys (and grids) can become stressed or traumatised, either by influence of underground cavities or mineral deposits, or from contamination by human activity such as quarries, mines and other industrial processes. In my own experience, the Curry grid is more affected by stress in the Earth, and is often found to be a contributory factor in water-related geopathic stress situations; whereas the Hartmann seems more influenced by human emotional trauma and geomantically-placed structures. Both grids can be good indicators of the presence of geopathic stress. When a building is suffering from geopathic stress the grid lines will move closer together into a tighter weave; when the stress is cleared the grids will gradually return to normal.

Unlike water lines, energy leys and grids have a much higher informational capacity; thus we find that they can carry human emotion, intent and even cognitive thought patterns. Much more research needs to be done in this area, but there is some evidence to show that moments of intense emotional trauma such as murders, fires, road accidents, magical rituals and the like can be 'imprinted' onto a site, making further incidents of a similar nature more likely to happen in the same place. Energetic repercussions of such can be transmitted along energy leys, inducing a similar resonance at other points on the ley. There is an excellent little book by Robin Heath (*Powerpoints –Secret*

Rulers and Hidden Forces in the Landscape) that investigates this area in more depth[39].

Human discarnates and other spirit energies can also propagate along energy leys, and incidents of haunting and poltergeist activity on leys are well known. Treatment and balancing of traumatised energy leys is possible, but this sort of healing is an advanced technique and beyond the scope of this beginners' guide. Novice dowsers should not attempt work of this sort without further training as there can be psycho-spiritual repercussions for the unwary.

It's a conscious Universe out there!

Why so many grids? Can all these energies really all exist side by side, overlapping and intermingling? Sometimes it seems as though a dowser only has to think about a grid to be able to find one. Is it all in the mind? Are we dowsing different frequencies or wavelengths of the same energies that we know as energy leys, or something else entirely - possibly the warp and weft of the geomagnetic field, telluric radiation, perturbations in the ether, bubbles in the quantum substructure of the universe, or even eddies in the space-time continuum (*'Ah,'* nodded Arthur, *'is he indeed?'*)?[40]

Dowsing Magic

Or could it be that we are dealing with phenomena at the very limits of human comprehension, a quantum world where language and conventional understanding fail us and our consciousness somehow imposes a grid structure upon our findings in order that we can make any sense of it at all? American researcher William Buehler seems to think so: "A grid is any system that organizes consciousness and energy patterns that support the consciousness and its intention."[41]

There are some other energy forms that are worth mentioning here. As Hamish Miller has demonstrated, you can consciously interact with a power centre and it will respond with a dowsing manifestation of a flower-like pattern.[42] Placing a large crystal on the power centre will have a similar effect.

Experiments have also been carried out in agricultural circles where geometric patterns like pentagons and 'Golden Eggs' have been constructed using nothing more than string, wooden pegs and some focused intent; and small but definite rises in temperature have been recorded within the shape. This technique was used in the Isles of Scilly to bring tulip crops into bloom earlier than normal in order to meet the Christmas market.[43].

This is where we begin to leave the rational scientific world behind, and enter into something like Tielhard de Chardin's *nöosphere* – a magical Universe more familiar to shamans, witches and magicians, where thoughts have power, otherworldly intelligences and entities exist, information about far-off places or events can be found remotely, and consciously directed *intent* can produce tangible changes in the physical world. As explained in Chapter One, dowsing lets us safely engage with these realms by enabling us to interrogate our subconscious whilst keeping one metaphorical foot in the real world; the pendulum acts as both filter and protective boundary and is an ideal tool to have along with you as you explore your particular spiritual path, whatever that may be. I think this is where we are going to see exciting developments in dowsing in years to come.

Conclusion

We have come a long way on our journey of exploration. We started out by trying to find water, and now here we are using our dowsing to ask questions about the very nature of reality. These questions also occupy the minds of our top scientific thinkers today, and yet as scientist and Dowsing Research Group member Jeffrey Keen says in the preface to his excellent book: "Conventional science cannot explain many phenomena in nature that seem to involve concepts such as non-locality or coherence, all of which imply instantaneous communication across vast distances in space-time."

Conventional science can't explain these things – but perhaps dowsing can? Jeffrey continues; "Dowsing… is a very powerful tool in unlocking the mysteries of non-locality, consciousness, and the nature of the Universe."[44]

Dowsing has that effect – used responsibly, it really does 'open doors in your mind'. All you have to do is step through, follow the path, and embrace the adventure. Your world will never be the same again.

Notes and References

1 'Stone ring' is the more politically-correct term for stone circle these days, as relatively few stone circles are actually circular.

2 Sir William Barrett & Theodore Besterman: *The Divining Rod: An Experimental and Psychological Investigation* 1926

3 The ideo-motor response (or "ideo-motor reflex"), often abbreviated to IMR, is a concept in hypnosis and psychological research. It is derived from the terms 'ideo' (idea, or mental representation) and 'motor' (muscular action). The phrase is most commonly used in reference to the process whereby a thought or mental image brings about a seemingly "reflexive" or automatic muscular reaction, often of minuscule degree, and potentially outside of the awareness of the subject. [13]

4 Sir William Barrett & Theodore Besterman: The Divining Rod: An Experimental and *Psychological Investigation* 1926

5 Dean Radin: *Entangled Minds* 2006. People with low latent inhibition tend to be more creative and open to new experiences; traits that are common in many dowsers.

6 http://en.wikipedia.org/wiki/Electromagnetic_theories_of_consciousness

7 Dr. Edith Jurka, *Brain Characteristics of Dowsers*, Journal of the British Society of Dowsers, Dec. 1983

8 Christopher Bird, *The Divining Hand*, Schiffer 1993

9 Uri Geller was purportedly paid one million dollars to dowse for oil deposits in South America.

10 Christopher Bird, *The Divining Hand*, Schiffer 1993

11 http://www.longrangelocators.com/forums/showthread.php?t=13032

12 Christopher Bird, *The Divining Hand*, Schiffer 1993

13 Slifierz, Richard: 'Dowsing at Gallipoli', *Dowsing Today,* September 2006

14 *Pennant* – the Journal of the Forces Pensions Society', November 2006

15 Christopher Bird, *The Divining Hand*, Schiffer 1993

16 Callers claiming to have located her through their dowsing were referred to the Crimestoppers phone line. Unless the BSD is approached in a formal capacity by some endorsed representative requesting the assistance of a dowser, this is not something that the Society can take an official stance on.

17 The Witchcraft Act was repealed in 1951 and replaced by the Fraudulent Mediums Act, wherein only those purporting to have psychic abilities for fraudulent purposes can be convicted. The very last person to be charged under the Act was 72-year-old East London medium Jane Rebecca Yorke in 1944.

18 Extract from BSD Code of Ethical Conduct: "Only dowse for information that concerns you personally or that lies within an area of public concern, unless you are asked or given permission by other people to dowse either for them personally or for groups or organisations of which they are members. Do not dowse for information about other people or their concerns without their permission, unless it is clearly in the interest of the highest common good to do so, and do not make unsolicited comments about other people or their concerns based on your dowsing."

19 Christopher Bird, *The Divining Hand*, Schiffer 1993

20 *ibid*.

21 Some dowsers recommend that you have the map correctly oriented, so that map north corresponds with True North. Personally I think this is a 'shibboleth', as I find it works just as well for me no matter what way the map is facing.

22 *Dowsing with Wands* - C Fortlage & E Phillips.

23 Sig Lonegren's hypothesis number 1: "Even if they were trained by the same teacher, when dowsing for intangible targets (like the Earth Energies), it is quite probable that no two dowsers will ever find exactly the same thing."

24 Online versions of both the EEG newsletter archive and the Encyclopaedia of Dowsing Terms are available on the EEG website: http://www.britishdowsers.org/EEG_site/home.html

25 http://en.wikipedia.org/wiki/Ley_lines

26 Hence the reason why energy leys are often found corresponding to a visual ley.

27 You can find the UVG placemark at: http://www.vortexmaps.com/hagens-grid-google.php. There's also a link on the EEG website.

28 http://belinusline.com/

29 Thanks to Richard Creightmore for this insight.

30 http://www.britishdowsers.org/EEG_site/archive/articles/axa1998_issue_12/SOLSTICE_clean.htm

31 http://www.britishdowsers.org/EEG_site/archive/articles/avc2000_issue_18/Jim%20Lyons%20article_clean.htm

32 http://www.consumerhealth.org/articles/display.cfm?ID=19990303212216

33 http://www.sncmfg.com/telecom/stray_voltage/dahlberg.html

34 John Burke & Kaj Halberg: *Seed of Knowledge, Stone of Plenty* – Understanding the Lost Technology of the Ancient Megalith Builders, 2005

35 http://www.britishdowsers.org/EEG_site/archive/articles/arb2004_issue_35/gridlines_lyons.htm

36 Blanche Merz: *Points of Cosmic Energy*, 1995.

37 http://www.britishdowsers.org/EEG_site/Glossary/page2.shtml#energygrids

38 http://www.britishdowsers.org/EEG_site/archive/articles/aqb2005_issue39/schneck.htm

39 Robin Heath: *Powerpoints – Secret Rulers & Hidden Forces in the Landscape*. Bluestone Press 2005

40 A simply irresistible quote from Douglas Adams' Life, The Universe, and Everything.

41 http://www.britishdowsers.org/EEG_site/archive/articles/arc2004_issue34/GrailLines.htm

42 See Hamish Miller's *Spirit of the Serpent* (DVD) for an example.

43 A. P. Tabraham: *Solar Energy and Dowsing in the Isles of Scilly*, 1982

44 Jeffrey Keen: *Consciousness, Intent, and the Structure of the Universe*. Trafford 2005

Bibliography

Sir William Barrett & Theodore Besterman: *The Divining Rod: An Experimental and Psychological Investigation 1926,* Kessinger 2004. ISBN 1-4179-8266-7

Harry Bell, *Forgotten Footsteps*, Ley Line Publications 1977

Christopher Bird, *The Divining Hand*, Schiffer 1993. ISBN 0-924608-16-1

John Burke & Kaj Halberg: *Seed of Knowledge, Stone of Plenty – Understanding the Lost Technology of the Ancient Megalith Builders*, Council Oak Books 2005. ISBN 1-57178-184-6

Tom Graves: *The Diviner's Handbook*, Destiny Books 1987. ISBN 0-89281-303-2

Tom Graves: *Needles of Stone*, Granada 1980. ISBN 0-586-04965-7

Robin Heath: *Powerpoints – Secret Rulers & Hidden Forces in the Landscape*. Bluestone Press 2005 ISBN 0-9526151-3-4

Jeffrey Keen: *Consciousness, Intent, and the Structure of the Universe*. Trafford 2005. ISBN 1-41204512-6

Blanche Merz: *Points of Cosmic Energy*, C W Daniel Co. 1995. ISBN 0-85207-194-9

Hamish Miller & Paul Broadhurst: *The Sun & The Serpent,* Mythos 1990. ISBN 978-0951518311

Dean Radin: *Entangled Minds – Extrasensory Experiences in a Quantum Reality*. Paraview/Pocket 2006. ISBN 978-1-4165-1677-4

Dennis Wheatley: *Principles of Dowsing*, Thorsons 2000. ISBN 0-7225-4029-9

Resources

The British Society of Dowsers
4/5 Cygnet Centre
Worcester Road
Hanley Swan
WR8 0EA
Email: info@britishdowsers.org
Telephone: +441684 576969
www.britishdowsers.org

Canadian Society of Dowsers
P. O. Box 641
Kingston, Ontario K7L 4X1
Email: info@canadiandowsers.org
Telephone: +1647 932 8612
www.canadiandowsers.org

Canadian Society of Questers
P.O. Box 4873
Vancouver, B.C., V6B 4A6
Email: ccss@aebc.com
www.questers.ca

American Society of Dowsers
Post Office Box 24
184 Brainerd St.
Danville, VT 05828
U.S.A.
Email: asd@dowsers.org
www.dowsers.org